AF565091

Food Science Technology and Agricultural Products

(The Changing Scenario)

Food Science Technology
and
Agricultural Products

(The Changing Scenario)

Edited by

Prof. Dheer Singh
Dr. D.K. Bhatt

SHREE PUBLISHERS & DISTRIBUTORS
NEW DELHI-110 002

Edition : 2013

Published by :
SHREE PUBLISHERS & DISTRIBUTORS
22/4735 Prakash Deep Building,
Ansari Road, Darya Ganj,
New Delhi–110 002

ISBN : 978–81–8329–505–5

Printed by :
Anvi Composers
Delhi

Preface

Food industry is one of the most emerging industries in the present scenario. As such the food processing industry is one of the largest in India – ranked fifth in terms of production, consumption, export and expected growth. The food industry is on a high as Indians continue to have a feast. The demand for food and specially the processed food is increasing day by day and there is need for development of new products and research in the field of food technology.

The present book is a collection of different researches done by eminent researchers in the field of food technology from the world over. The papers being presented in this volume were presented at the National Conference on Changing Scenario of Food Science Technology and Agricultural Products. The chapters include Bakasang Processing (a product of Malaysia), Effect of Artificial Neural Networks For Forecasting Shelf Life of Paneer Tikka, Effect of Guar Gum Hydrolysis on Quality of Cookies, New Techniques in Rice Bran Oil Processing, Nutritional And Health Benefits of Soybeans And Micro Enterprise Opportunities, Changing Scenario of Food Processing Industry, Influence of Optimal Harvest Stage And After Ripening on Seed Quality of Processing of Type Hot Pepper, Prediction of Cookie Quality of Indian Wheat Varieties by Solvent Retention Capacity Profile, Food Quality Control And Food Safety Management, Dehydration of Vegetables By Application of Radio Frequency Heating, Nutritional Status of Barseem in India besides many other new topics of interest for the students and researchers in the field of food technology.

The book will be of maximum use for all the researchers in the field of food science, related technology and agriculture. It is aimed that the book will fill the lacuna in the published work in food technology and will act as a food for thought for further research in the various streams of food science and technology.

We would like to take this opportunity to extend our gratitude to all the contributors of this book, and also to all the staff and faculty member of the Institute of Food Technology, Bundelkhand University, Jhansi for their support and cooperation.

Prof. Dheer Singh

Dr. D.K. Bhatt

Contributors

A.S. Bawa Director, Defence Food Research Laboratory, Siddharthanagar, Mysore, Karnataka-570011.

Anubha Shukla School of Biotechnology, Gautam Buddha University, Greater Noida, U.P.–201308.

B.L. Lokeshappa Agriculture Officer, Dept. of Agriculture, Karnataka.

B.S. Khatkar Department of Food Technology, Guru Jambheshwar University of Science and Technology, Hisar, Haryana.

Charli Kaushal Gautam Buddha University, Greater Noida.

D.K. Bhatt Institute of Food Tech., Bundelkhand University, Jhansi.

D.T. Khogare Subject Matter Specialist (Home Science), Krishi Vigyan Kendra, Sangli. (MS) India.

Deepak Mudgil Department of Food Technology, Guru Jambheshwar University of Science and Technology, Hisar, Haryana.

Dheer Singh Director, Institute of Food Technology, Bundelkhand University, Jhansi (UP)–284 128.

G.K. Goyal Principal Scientist, Dairy Technology Division, National Dairy Research Institute, Karnal.

H.K. Sharma Dean, Research of Consultancy, Sant Longowal Institute of Engineering and Technology (Deemed University), Longowal-148106, Sangrur, Punjab.

Kalyan P. Babar Post Graduate Institute, Mahatma Phule Krishi Vidyapeet, Rahuri. Distt-Ahmednagar (Maharashtra).

Kunal K. Ahuja National Dairy Research Institute, Karnal.

K. Radhakrishna Defence Food Research Laboratory Siddathanagar, Mysore-570011.

K. Shukla Department of Biotechnology, Gautam Buddha University.

L. Ghorbani Dept. of Agricultural Machinery, Shahid Bahonar University of Kerman, Jomhourieslami boulevard, Kerman, Iran.

M. Shamsi Department of Agricultural Machinery, Shahid Bahonar University of Kerman, Jomhourieslami boulevard, Kerman, Iran.

M.C. Pandey Defence Food Research Laboratory, Siddharthanagar, Mysore, Karnataka.

Mukul Sinha Assoc. Prof.-cum Sr. Scientist, Deptt. of Food & Nutrition, College of Home Science, RAU, Pusa.

Nurul Huda	Fish and Meat Processing Laboratory, Food Technology Programme, School of Industrial Technology, University Sains Malaysia.
P.N. Shastri	Dept. of Food Tech., Laxminarayan Institute of Technology, R.T.M. Nagpur University, Nagpur. (MS).
Prabal Talukdar	Department of Mechanical Engineering, Indian Institute of Technology Delhi, New Delhi.
Preety Pant	School of Biotechnology, Gautam Buddha University, Greater Noida, Uttar Pradesh– 201308.
P.T. Harilal	Defence Food Research Laboratory, Siddharthanagar, Mysore, Karnataka-570011.
R.S. Gaikwad	Post Graduate Institute, Mahatma Phule Krishi Vidyapeet, Rahuri. Distt. Ahmednagar. (Maharashtra) 413722.
R. Ahirwar	Defence food research laboratory Siddathanagar, Mysore, India-570011
R. C. Mathad	Professor, Dept. of Seed Science and Technology, UAS, Raichur.
Ranjeet Varma	Faculty of Engi. and Tech., RBS College, Bichpuri, Agra.
S. Rustagi	Gautam Buddha University, Greater Noida, Uttar Pradesh.
S.B. Patil	College of Agriculture, Bhimarayana- gudi, UAS, Raichur.
Shalini Shrivastav	Chemistry Department, Eshan College of Engineering, Farah, Mathura.
S.N. Vasudevan	Professor & Head, Dept. of Seed Science and Technology, UAS, Raichur.
Shanky Bhat	School of Biotechnology, Gautam Buddha University, Greater Noida.
Sheweta Barak	Department of Food Technology, Guru Jambheshwar University of Science and Technology, Hisar.
S. Priyadarshi	School of Biotechnology, Gautam Buddha University, Greater Noida, U.P.
Sumit Goyal	National Dairy Research Institute, Karnal.
Suresh Itapu	NutriTech Consulting Services Pvt. Ltd., New Delhi.
S. Mathur	Gautam Buddha University, Greater Noida, Uttar Pradesh.
S. Shukla	School of Biotechnology, Gautam Buddha University, Greater Noida.
S. Yadav	Gautam Buddha University, Greater Noida, Uttar Pradesh.
V.P.C. Mohan	Department of Mechanical Engineering, Indian Institute of Technology Delhi, New Delhi.
VPS Bhadauria	Chemistry Department, Eshan College of Engineering, Farah, Mathura.
YVS Vynavi	GITAM University, Rushikonda, Vishakapatnam–500045.

Contents

Processing, Chemical Composition and Microbial Characteristics of Bakasang

Nurul Huda

Abstract

Indonesia is surrounded by waters that serve as source of high marine fish production. Traditional preservation and processing methods, such as salting, boiling, smoking and fermentation, are responsible for 54.67% of the total preserved and processed fish products. Bakasang is one of the traditional fermented fish product that is produced in North Sulawesi province and Moluccas Island. Physically, bakasang is a dark brown liquid product with a strong fishy flavor. It is usually used as a flavoring agent for many dishes or mixed with red chilies, tomatoes, red onions, and garlics, which are then sautéed with coconut oil and eaten with hot porridge mixtures of rice and vegetables called tinutuan. The main raw materials for bakasang processing are the internal organs (viscera and roe) of tuna and tuna-like fish obtained as a waste product from the processing of smoked tuna. This article will elaborate on the general processing methods of bakasang and present some data in relation to their chemical composition and microbial characteristics.

Introduction

Fermented fish products are important traditional foods in many countries of the world, particularly in Southeast Asia where fermented small fish, shrimp and intestines are widely consumed (Ishinge, 1993). The use of fermented fish products is widespread in Indonesia, Malaysia, Burma, Philippines and Thailand. In the past, fermented fish products were largely produced by households for consumption. However, the pattern on production, distribution, and consumption is changing rapidly owing to population growth and increased urbanization.

Fish sauce, one of the popular fermented fish product, is extensively used as condiment in many countries of Southeast Asia, especially as supplements in the diets of the poorer section of the population. Fish sauce is also prepared and consumed in the Shantung peninsula of northern China and in parts of Japan (Ishinge, 1993). Fish sauce is known as budu in Malaysia and Thailand (Karim, 1993; Phithakol, 1993), kecap ikan, rusip and bakasang in Indonesia (Putro, 1993), nga ngan-pya-ye in Burma (Tyn, 1993) and patis in Philippines (Mabesa and Babaan, 1993).

Indonesian Fermented Fish Product

Indonesia consists of 17,480 islands and surrounded with waters that provide a high production of fresh water and marine fish. Fish is popular food product among Indonesian and fish consumption (kg) per person increased from 23.95 in 2005 to 29.08 in 2009. Total production of fish at 2005 was 6,869,543 tons and increased to 10,835,618 tons in 2010. Fish production is dominated by marine fish which contributed around 78% of the total fish produced in 2010 (Departemen Kelautan dan Perikanan, 2011).

Fish is mainly consumed in the fresh form, and the rests (around 40%) are subjected to different preservation and processing methods. Traditional preservation and processing methods, such as salting, boiling, smoking and fermentation, are responsible for 55% of the total preserved and processed fish products. The rest (45%) of fisheries production are subjected to freezing, canning and fish meal production. Among the traditional preservation and processing methods, fermentation contributes less than 2% of the total preserved and processed fish products. Terasi, peda and fish sauce is the common fermented fish product produced in Indonesia. Fermented terasi is the main fermented fish product available in Indonesia followed by fermented peda. Fish sauces such as kecap ikan, rusip and bekasang are produced in small amounts. Table 1.1 shows some data of the Indonesian marine capture fisheries production by type of disposition.

TABLE 1.1

Indonesian Marine Capture Fisheries Production by type of Traditional Preservation and Processing Methods Disposition

(tons)

Disposition	*2005*	*2007*	*2009*
Drying and Salting	716392	717240	637026
Boiling	124463	164534	142838
Fermentation Terasi	38026	19915	25243
Fermentation Peda	9137	16556	15435
Fermentation Fish Sauce	238	748	577
Smoking	115200	104972	121266
Others	49482	40423	32678

Fish Sauce

Fish sauce is normally referred to as kecap ikan. *Kecap ikan* or fish sauce is a clear yellow to brown liquid with a salty taste and specific aroma. In Indonesia, *kecap ikan* is the third largest fermented fish product, produced and concentrated along Java Island. However other types of fish sauce such as rusip and bakasang are also produced in other parts of Indonesia. Rusip is produced and consumed among the people of South Sumatera, while bakasang is produced and consumed among the people of North Sulawesi and Moluccas Island. Ijong and Otha (1995) clearly stated that bakasang is another form of fish sauce. Studies on rusip and bakasang have been very limited and information on their processing, chemical composition and microbial characteristics is very scarce.

Bakasang, Another Type of Fish Sauce

Bakasang is a dark brown liquid product with a strong fishy flavor (Figure 1.1). *Bakasang* is a fermented fish product mainly produced in North Sulawesi

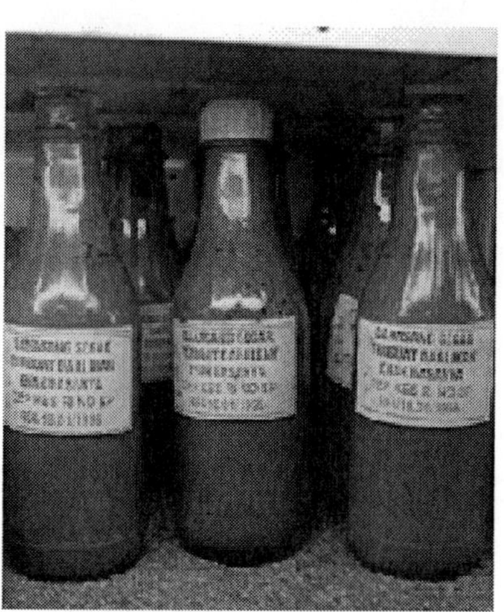

Figure 1.1: *Bakasang and Tinutuan*

province and Moluccas Island. It is usually used as a flavoring agent for many dishes or mixed with red chilies, tomatoes, red onions, and garlics, which are then sautéed with coconut oil and eaten with hot porridge mixed with rice and vegetables called *tinutuan* (Ijong and Ohta, 1995). The main raw materials for *bakasang* processing are the internal organs (viscera and roe) of tuna and tuna-like fish obtained as a waste product from the processing of smoked skipjack tuna (*Katsuwonus pelamis*). In addition to the internal organs of tuna, *bakasang* is also produced from the whole body of small marine fish. Smoked skipjack tuna is a popular product in the North Sulawesi province and the Moluccas Island (Subroto, 1985). Subroto and others (1984) reported that *bakasang* made from ground viscera was of better quality compared to those from whole viscera. The internal organs of yellowfin tuna (*Thunnus macropterus*) and big eye tuna (*Thunnus obesus*) are also good raw materials for the processing of *bakasang* (Poernomo, 1996). Ijong and Ohta (1996) and Harikedua and others (2009) reported that some small marine fishes, such as sardine (*Sardinella* sp.), anchovy (*Stolepherus* sp. or *Engraulis* sp.) and orange sunkist shrimp (*Caridina wycki*), are also suitable for *bakasang* processing.

Bakasang Processing Method

The traditional processing of *bakasang* starts with the collection of internal organs, including the viscera and roe of skipjack tuna (Figure 1.2). An appropriate amount of the internal organs are washed with seawater and drained to remove excessive seawater. The cleaned internal organs are mixed with 15-20% salt and poured into bottles. The bottles are kept for 3–6 weeks to allow fermentation to occur until a dark and sticky liquid is formed. Traditionally, the bottles are stored in a warm place and usually near a source of fire. Subroto and others (1984) stated that *bakasang* could be made from the visceral parts of bonito by adding 10% salt and allowing the mixture to ferment for 10 days.

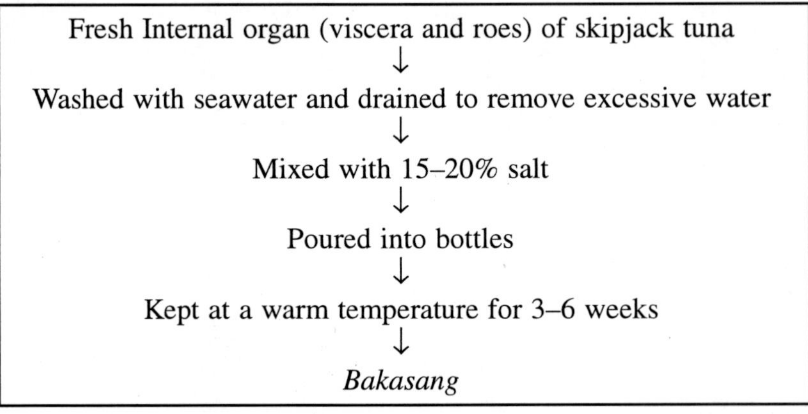

Figure 1.2: *General Procedure for Bakasang Processing*

Improved method of bakasang processing was suggested by Ijong and Otha (1995c) and Ijong and Otha (2000). Bakasang prepared by using LAB-mixed culture will improve the quality of the final product by inhibiting the extent of rancidity and increasing the protein content. Sensory evaluation showed that addition of up to 5% LAB-mixed culture is quite acceptable. Addition of LAB-mixed culture successfully increased the content of certain amino acids such as aspartic acid, glutamic acid, praline and valine. These amino acids contribute to the preferred flavor of bakasang. Up dated report by Ijong and Otha (2000) showed that addition of single culture of *Lactobacillus* strain T38 was able to produce better quality of bakasang than LAB-mixed culture. Extent of rancidity of bakasang with *Lactobacillus* strain T38 is lower and the sensory score for flavor, aroma and color was higher compared to bakasang without the addition of *Lactobacillus* strain T38.

Chemical Composition

As a liquid food product, *bakasang* contains high moisture content, which is approximately 66–69% (Ijong and Ohta 1995a). Harikedua and others (2009) also reported the moisture content of *bakasang* to be approximately 59–70%. The salt content of bakasang was influénced by the amount of salt used during *bakasang* preparation. Previously, Ijong and Ohta (1995) reported that the salt content of *bakasang* was approximately 84–180 g/kg sample; however, report by Harikedua and others (2006) reported that the salt content of *bakasang* was 78–98 g/kg sample. Chemical composition of *bakasang* is presented in Table 1.2.

TABLE 1.2

Chemical Composition of *Bakasang*

Parameter	*Ijong and Otha (1995a) - Commercial*	*Ijong and Otha (1995b) - Experimental*	*Harikedua and others (2006) – Commercial*
Moisture	66.3 – 68.9	66.9 – 71.9	59.0 – 70.9
Protein (% wb)	14.0 – 17.4	15.2 – 16.7	-
Fat (% wb)	1.0 – 3.0	1.6	-
Ash (% wb)	-	-	-
Salt (% wb)	8.4 – 18.0	8.9 – 15.8	7.6 – 9.8
pH	5.8 – 6.3	5.5 – 5.8	6.4 – 6.5
Acidity (mL/L NaOH)	21.9	22.2 – 25.7	-
TVN (mg/100 g)	-	-	345.7 – 351.4
Titratable Acid (%)	-	-	1.2 – .1.9
Free Amino Acid Nitrogen (mg N/mL)	98.9	104.5– 131.3	802.2 – 921.6

A quality *bakasang* product is related to a high moisture content, low viscosity, low titratable acid, low salt content and low free amino nitrogen content (Harikedua and others 2006). A longer fermentation process will produce *bakasang* with higher free amino nitrogen content, which is associated with lower consumer acceptability. Bakasang with free amino nitrogen content of 802.19 mgN/mL is more preferred than bakasang with free amino nitrogen content of 921.56 mgN/mL.

Microbial Characteristics

In terms of microbial quality, Ijong and Ohta (1995) reported that *Staphylococcus* sp. and *Lactobacillus* sp. were the predominant bacteria isolated from *bakasang*. Other bacteria isolated in *bakasang* are *Micrococcus* sp., *Streptococcus* sp., *Bacillus* sp., and *Clostridium* sp. *Bakasang* prepared from small sardines (*Engraulis* sp.) contained additional bacteria such as *Enterobacter* sp., *Moraxella* sp., and *Pediococcus* sp. (Ijong and Ohta, 1996). Microbial quality of commercial bakasang presented in Table 1.3.

Vernam and Evans (991) and Ijong and Otha (1995) stated the presence of Lactobacillus in the fermentation of bakasang could contribute to lowering the pH and preventing the growth of undesirable organisms such as coliform or pathogenic bacteria. Salt also acts as bacteriostatic for most of the bacteria. Generally known, the use of salt in fermentation is to prevent the growth of spoilage bacteria, including the pathogenic types. The presence of *S. aureus* and *Clostridium* in commercial sample seems to be attributable to the contamination of raw materials used during processing. However, this kind of contamination can be prevented during processing if the processing is done under hygienic conditions.

TABLE 1.3

Microbial Quality of Commercial Bakasang (log cfu/mL) (Ijong and Otha, 1995; Ijong, 2010).

Parameter	*Sample A*	*Sample B*	*Sample C*	*Sample D*
TPC aerobic	5.72	6.08	5.44	5.03
TPC anaerobic	3.38	1.96	3.84	3.27
LAB	6.15	4.95	4.99	4.97
Coliform	ND	ND	ND	ND
SFB	ND	ND	2.08	5.01
Staphylococcal	6.38	5.99	5.34	4.72
Microflora	*Staphylococcus*	*Staphylococcus*	*Staphylococcus*	*Staphylococcus*
Isolated	*S. aureus,* *S.epidermidis* *Lactobacillus*	*S. aureus* *Micrococus* *Streptococcus*	*S. aureus* *Lactobacillus* *Streptococcus* *Clostridium*	*S. aureus* *S.epidermidis* *Lactobacillus* *Clostridium* *Bacillus*

Conclusion

Fermented fish product is one of the traditional food product produced and consumed among Indonesian people. Bakasang, is one of unique fermented fish product which popular among North Sulawesi province and Moluccas Island people. Chemical composition and microbiological quality of bakasang is different and influenced by many factors such as raw material, ingredient used and level of hygienic condition applied.

References

Departemen Kelautan dan Perikanan, 2010. Statistik perikanan tangkap. Jakarta: Direktorat Jenderal Perikanan Tangkap Departemen Kelautan dan Perikanan Indonesia.

Harikedua, S.D., Wijaya, C.H., and Adawiyah D.R. 2009. Keterkaitan mutu fisiko-kimia dengan atribut sensori :"Bakasang". Paper presented at Seminar Nasional PATPI, Jakarta 3-4 November 2009.

Ijong, F.G., and Ohta, Y. 1995a. *Microflora and Chemical Assessment of an Indonesian Traditional Fermented Fish Sauce "Bakasang"*. J Fac Appl Biol Sci Hiroshima Univ. 34:95-100.

Ijong, F.G., and Ohta, Y. 1995b. *Amino Acid Compositions of Bakasang*, A Traditional Fermented Fish Sauce from Indonesia. LWT 28:236-237.

Ijong, F.G., and Ohta, Y. 1995c. Characteristics of Bakasang Fermented with Lactic Acid-mixed Culture. J Fac Appl Biol Sci Hiroshima Univ 34:1-10..

Ijong, F.G., and Ohta, Y. 1996. Physicochemical and Microbiological Changes associated with Bakasang Processing - A Traditional Indonesian fermented Fish Sauce. J Sci Food Agric 71:69-74.

Ijong, F.G., and Ohta, Y. 2000. *Utilization of Lactic Acid Bacteria in Bakasang Production using a Semi-continous Fermentor*. In Proceeding of The JSPS-DGHE International Symposium on Fisheries Science in Tropical Area. Bogor August 21–25. Pp. 462-466.

Ijong, F.G. 2010. Karakteristik fisik-kimiawi dan mikrobiologis produk fermentasi ikan "bakasang" dari Sulawesi Utara. Prosiding Seminar Nasional Pangan, 2010.

Ishinge, N. 1993. *Cultural Aspects of Fermented Fish Products in Asia.* In: **Lee C.H., Steinkraus K.H., Reilly PJA**, editors. *Fish Fermentation Technology.* Tokyo: United Nations University Press. p 13-32.

Karim, M.I.A. 1993. *Fermented Fish Products in Malaysia.* In. **Lee C.H., Steinkraus K.H., Reilly PJA**. Editors. *Fish Fermentation Technology.* Tokyo: United Nations University Press. p 95-106.

Mabesa, R.C. and J.S. Babaan. 1993. *Fish Fermentation Technology in the Philippines.* In. Lee CH, Steinkraus KH, Reilly PJA. Editors. *Fish Fermentation Technology.* Tokyo: United Nations University Press. p 85-94.

Phitakol, B. 1993. *Fish Fermentation Technology in Thailand.* In. **Lee C.H., Steinkraus K.H., Reilly PJA**. Editors. *Fish Fermentation Technology.* Tokyo: United Nations University Press. p 155-166.

Poernomo, J. 1996. Pengaruh tapioka dan garam dalam fermentasi bakteri asam laktat jeroan ikan tuna (Thunnus sp.). Buletin Teknologi Hasil Perikanan II(2):64-73.

Putro, S. 1993. *Fish Fermentation Technology in Indonesia.* In: Lee CH, Steinkraus KH, Reilly PJA, editors. *Fish Fermentation Technology.* Tokyo: United Nations University Press. p. 107-128.

Subroto, W., Setiabudi, E, and Sjahrul, B. 1984. *Preliminary Study on Production of Bakasang.* Laporan Penelitian Teknologi Perikanan 26: 9-17.

Tyn, M.T. 1993. *Fermented Fish Products in Burma.* In. **Lee C.H., Steinkraus K.H., Reilly PJA**. Editors. *Fish Fermentation Technology.* Tokyo: United Nations University Press. p. 129-154.

Vernam, A.H., and Evans, M.G. 1991. *Foodborne Pathogens: An Illustrated Text.* Toronto, Mosby year book.

○○

2

A Scientific Study on Artificial Neural Networks for Forecasting Shelf Life of Paneer Tikka

Kunal K. Ahuja, Sumit Goyal, D.K. Bhatt and G.K. Goyal

Abstract

'Paneer tikka' is an exotic kebab of Indian cottage cheese, which is highly rich in proteins, vitamins, minerals, fibre content and sulphur compounds. Typically, the shelf life of 'paneer tikka' is hardly one day at ambient temperature. In this study, the product was vacuum packed in two high barrier bags, viz., *nylon based co-extruded films (P1) and metallised polyester films (P2) along with Low Density Polyethylene (P3) as control, which were stored at 3±1°C so as to enhance its shelf life. The stored samples were evaluated at regular intervals for changes in physico-chemical characteristics,* viz., *moisture, titratable acidity, free fatty acid and tyrosine content. However, flavour scores and overall acceptability scores were obtained through a panel of trained sensory judges. The shelf life of vacuum packed 'paneer tikka' samples significantly increased up to 40 days for the product packed in P1 and P2, while the samples packed in P3 were acceptable only up to 20 days. Further, a connectionist model has been proposed to predict sensory scores of 'paneer tikka' using the sample data. The back-propagation learning algorithm with Bayesian regularisation mechanism was used. The connectionist model with two hidden layers having five neurons in first hidden layer and three neurons in second hidden layer was found to be the best configuration to predict flavour score and overall acceptability score with root mean squared error as 7.27% and 4.71%, respectively. Finally, the residual shelf life of the 'paneer tikka' was computed using regression equations based on flavour and overall acceptability scores.*

Keywords: *ANN model, Paneer tikka, Shelf life, Forecasting.*

Introduction

'Paneer' is an important acid coagulated indigenous variety of soft cheese, which forms base for a variety of culinary dishes, stuffing material for various vegetable dishes, snacks, and sweetmeats. 'Tikka' is a general South Asian term meaning marinated barbequed food. 'Paneer Tikka' is a very popular dish among vegetarians, which is highly rich in proteins, vitamins, minerals, fibre content and sulphur compounds; and is laxative in nature. According to Sinclair (2004), 'tikka' means small pieces of meat or poultry marinated in yoghurt and tandoori spice mix, grilled or roasted and served with Indian bread and salad. It is extensively used as fast-food during get-togethers, marriage parties, birthday parties, and also in restaurants. 'Paneer tikka' requires more than five hours for its preparation. At present 'paneer tikka' is consumed immediately after its preparation; however, if stored, its shelf life is hardly one day, which impedes its proper marketing and distribution. To impart longer shelf life, the product was vacuum packed in two high barrier bags, *viz.*, nylon based co-extruded films (P1) and metallised polyester films (P2) along with Low Density Polyethylene (P3) as control and stored at 3±1°C. Awareness of modern consumers about quality of the processed foods creates challenge in product development and marketing. Shelf life prediction models could facilitate the manufacturers and also the food quality regulating agencies to critically evaluate the shelf life of such foods.

An Artificial Neural Network (ANN), usually called Neural Network (NN), is a mathematical model or computational model that is inspired by the structure and/or functional aspects of biological neural networks. A neural network consists of an interconnected group of artificial neurons, and it processes information using a connectionist approach to computation. In most cases an ANN is an adaptive system that changes its structure based on external or internal information that flows through the network during the learning phase. Modern neural networks are non-linear statistical data modelling tools. They are usually used to model complex relationships between inputs and outputs or to find patterns in data (Wikipedia website, 2011).Some studies showing the use of connectionist models for predicting quality/shelf life of various food products have been reviewed, *e.g.*, Vallejo *et al.* (1995) studied the usefulness of ANN models *vis-à-vis* Principal Components Regression (PCR) model for predicting shelf life of milk and found better predictability of ANN model than the PCR model; Ni and Gunasekaran (1998) observed that a three-layer ANN model was able to predict the rheological properties of Swiss type cheeses on the basis of their composition more accurately than regression equations; **Sofu and Ekinci (2007)** developed an ANN model using back-propagation learning algorithm for prediction of shelf life of set-type whole-fat and low-fat yoghurts, and found that the ANN model was able to analyse nonlinear multivariate data with very good performance, fewer parameters and shorter calculation time; Goñi *et al.* (2008) trained and validated a feedforward ANN model using experimental values of freezing and thawing times of foods and test substances of different geometries, and found that ANN model provided a simple and accurate prediction method for freezing and thawing times; Singh *et al.,* (2009) studied the sensory quality of Ultra High Temperature (UHT) processed milk using the kinetic *vis-à-vis* ANN models, and observed that the prediction performance of the ANN models were found to be better than the kinetic models and out of the different ANN approaches examined, the Bayesian regularisation algorithm provided the most consistent results; and Ruhil *et al.* (2009) examined ANN approach *vis-à-vis* regression models based on the predictive sensory quality data to predict the shelf life of 'basundi mix' in terms of residual storage life. In the light of above review of literature, this study was undertaken to forecast the shelf life of 'paneer tikka' using suitable ANN model based on backpropagation learning algorithm with Bayesian regularisation as training function.

Experiments, Results and Discussion

Dataset

The experimental data on quality parameters, *viz.,* moisture, titratable acidity, free fatty acid, tyrosine content, pH and water activity were obtained as a result of storage analysis of 'paneer tikka'. However, the parameters, which were quick to obtain as well as statistically showed their higher correlation with sensory parameters, were selected as input parameters. Although, water activity exhibited a higher correlation (*i.e.*, 0.95 with flavour score and 0.95 with overall acceptability score) and was quick to obtain, but no significant difference ($P<0.05$) was found between intervals of storage. Thus, water activity was excluded from the dataset. Finally, moisture, titratable acidity, free fatty acids and tyrosine content were selected as input parameters. The sensory score for flavour and overall acceptability were used as output parameters for building the model. A five-point hedonic scale was used for the evaluation of sensory parameters. Experimentally determined 42 observations for each input and output parameters were considered for the modelling. The dataset was randomly divided into two disjoint subsets, namely, training set containing 34 observations (*i.e.*, 80% of total observations) and test set comprising of 8 observations (*i.e.*, 20% of total observations).

ANN Model

In the development of ANN model for shelf life prediction of 'paneer tikka', different combinations of several internal parameters, *i.e.*, data pre-processing, data partitioning approach, number of hidden layers, number of neurons in each hidden layer, transfer function, error goal, *etc.*, along with back-propagation algorithm based on Bayesian regularisation mechanism as training function, were empirically explored in order to optimise the prediction ability of the model. It is evident from literature, that there is no generalised method to determine the optimum values for these parameters, *i.e.*, number of hidden layers and neurons in each hidden layer, *etc.*, maybe as they are function of expected intelligence. Hence, the 'trial and error' approach was pursued to decide the optimum architectural parameters. Connectionist models with single hidden layer as well as with two hidden layers were explored. The number of neurons in each hidden layer was varied from 2 to 14. Weights and biases were randomly initialised. The network was trained with 150 epochs. The transfer function for each hidden layer was tangent sigmoid function while that for the output layer was linear function. The backpropagation method was used as training algorithm. After finding optimum values for ANN network and preparing training and test datasets, the network was trained with training set. The network was then simulated with the test set as well as with fresh input dataset to validate the proposed connectionist model. In backpropagation algorithm, for a given set of input to the network, the response to each neuron in the output layer was calculated and compared with corresponding desired output response. The errors associated with desired output response were adjusted in a way that will reduce these errors in each neuron from the output to the input layer. The main problem faced while constructing the connectionist model was over fitting, *i.e.*, the error with the training set comes out to be a very small value but when a new dataset is employed to the network, the error becomes larger (Mittal and Zhang, 2000). To overcome such a problem, a variant of the back-propagation method based on Bayesian regularisation technique was used, which determines the optimal regularisation parameters in an automated fashion (Mackay, 1992; Foresee and Hagan, 1997). The Neural Network Toolbox under MATLAB software was used to develop the connectionist models. The performance of connectionist model was evaluated using Root Mean Square Error (RMSE) technique as follows:

$$RMSE = \sqrt{\frac{1}{N}\left[\sum_{1}^{N}\left(\frac{Q_{\exp} - Q_{cal}}{Q_{\exp}}\right)^{2}\right]} \times 100$$

where, $Q_{\exp}$ = Observed value,

Q_{cal} = Predicted value and

N = Number of observations in dataset.

Performance matrices of the connectionist model for predicting flavour and overall acceptability sensory scores are presented in Table 2.1.

TABLE 2.1

Performance of ANN Model for Predicting Flavour Score and Overall Acceptability Score.

Number of hidden layers	*Number of neurons in hidden layer*		*RMSE*	
	I	*II*	*FS**	*OAS***
1	3	-	13.47	10.52
1	5	-	27.12	9.39
1	7	-	18.64	16.36
1	9	-	13.66	8.01
1	10	-	8.25	10.05

Number of hidden layers	*Number of neurons in hidden layer*		*RMSE*	
	I	*II*	*FS**	*OAS***
1	12	-	19.94	17.60
2	2	2	11.93	9.14
2	3	3	15.42	10.47
2	5	5	12.97	9.52
2	7	7	10.91	7.58
2	9	9	10.89	8.00
2	12	12	13.14	15.43
2	3	5	19.98	14.96
2	5	3	7.27	4.71
2	5	7	17.17	15.27
2	7	5	20.12	18.59

*FS-Flavour score, **OAS-Overall Acceptability Score

As the number of neurons were varied from 2 to 14 for hidden layers, the RMSE for flavour score and overall acceptability score were varied greatly from 7.27 to 27.12 and 4.71 to 18.59, respectively. The configuration for the best connectionist model to predict flavour score and overall acceptability score for 'paneer tikka' (Figure 2.1) included two hidden layers with five neurons in first hidden layer and three neurons in second hidden layer attaining RMSE as 7.27% for flavour score and 4.71% for overall acceptability score. Further increase in the number of neurons in each layer did not lead to any improvement in the network performance.

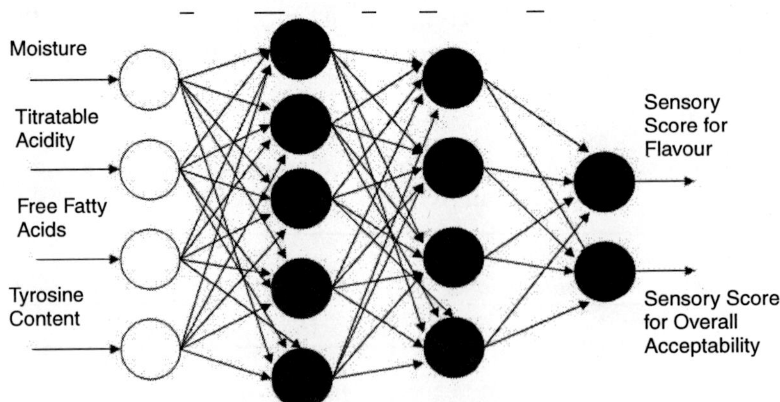

Figure 2.1: *ANN Model to Predict Flavour Score and Overall Acceptability Score for Paneer Tikka.*

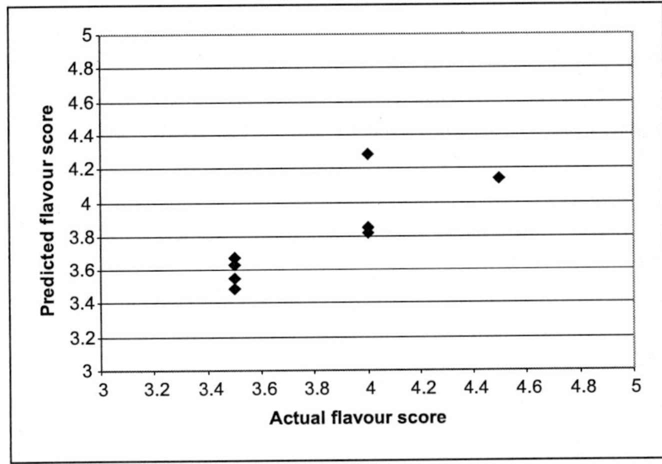

Figure 2.2: *Comparison of Actual and Predicted Flavour Score using ANN model.*

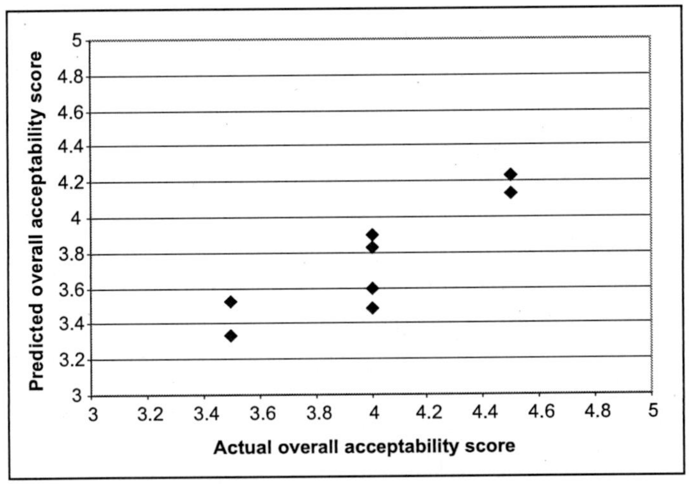

Figure 2.3: *Comparison of Actual and Predicted Flavour Score using ANN Model.*

The comparison between the actual flavour score and predicted flavour score using connectionist model is shown graphically in Figure 2.2 and that for overall acceptability score is shown in Figure 2.3.

Shelf Life Prediction

The regression equations (Table 2.2) were developed to estimate storage life of the product, *i.e.*, the time (*T*) for which the product has been in the shelf, based on either flavour score or overall acceptability score in known packaging material.

The residual life of the product was computed by first determining the shelf life of the product for the minimum acceptable flavour or overall acceptability score through experiments and then subtracting the computed storage life for predicted values of flavour or total sensory score through connectionist models. The resultant zero or less than zero value indicated that the product's shelf life was expired and was no regression equations were estimated for 'paneer tikka' vacuum

TABLE 2.2

Estimates of Parameters of Regression Equations

Packages	*Constant*	*Regression Coefficient*	*Coefficient of determination*
Flavour score			
P1	265.00	– 60.34	0.84
P2	227.89	– 51.20	0.87
Overall Acceptability Score			
P1	327.50	– 75.00	0.90
P2	201.84	– 45.92	0.83

packed in two different packaging materials, *viz.*, nylon based co-more suitable for consumption or for further storage. The extruded films (P1) and metallised polyester (P2) and stored at 3 ± 1°C taking storage intervals (in days) as dependent variable and sensory scores (flavour and overall acceptability) as independent variable. However, the regression equation for the samples packed in Low Density Polyethylene (P3) as control were not developed, because product was spoiled within 20 days of storage. All the equations were found to be the best fit and the value of coefficient of determination (R^2) ranged between 0.83 and 0.90 indicating that 83 to 90 percent of total variation was explained by flavour and overall acceptability scores. All the regression coefficients associated with the flavour and overall acceptability scores were found to be statistically significant ($P < 0.05$). For instance, the time period for which the product has been in shelf can be predicted for an arbitrary flavour score of 4.25 for the 'paneer tikka' samples packed under vacuum in packaging material P1 and stored at 3 ± 1°C.

$$T = 265.00 - 60.34 \times 4.25 = 8.55$$

Now, the residual shelf life (in days) is computed by subtracting the above obtained value of T from the experimentally computed shelf life, which was found to be 40 days in present case. Hence, the residual shelf life comes out to be 31.45 days. Similarly, the time period for which the product has been in shelf can be predicted on the basis of an arbitrary overall acceptability score of 3.5 for the samples packed in P2 and stored at 3±1°C.

$$T = 201.84 - 45.92 \times 3.5 = 41.12$$

Since, the value obtained above exceeds the experimentally computed shelf life of 40 days; therefore, the product should be discarded.

Conclusion

A feedforward backpropagation ANN model was developed with four input parameters, *viz.*, moisture, titratable acidity, free fatty acid and tyrosine content to predict flavour and overall acceptability scores for an Indian dairy product, 'paneer tikka'. The best ANN configuration for prediction of flavour score and overall acceptability score included two hidden layers with five neurons in first hidden layer and three neurons in second hidden layer, which achieved RMSE as 7.27% for flavour score and RMSE as 4.71% for overall acceptability score. The prediction performance of ANN models gave good fit. Further, the regression models were also developed for predicting the shelf life of 'paneer tikka' in terms of residual storage life in days. Hence, the outcome of this study allows us to infer that connectionist approach could be used as an effective tool for predicting the shelf life of dairy products for better quality control.

References

Http://en.wikipedia.org/wiki/Artificial_neural_network (Wikipedia website, accessed on 15.10.2011).

Forsee, F.D., Hagan, M.T., 1997. Gauss-Newton approximation with Bayesian regularization. In: Proceedings of the 1997 IEEE International Joint Conference on Neural Networks, vol. 3, pp. 1930-1935.

Goñi, S.M., Oddone, S., Segura, J.A., Mascheroni, R.H., Salvadori, V.O., 2008. *Prediction of Foods Freezing and Thawing Times: Artificial Neural Networks and Genetic Algorithm Approach.* Journal of Food Process Engineering 84(1), 164–178.

Mackay, D.J.C., 1992. *Bayesian Interpolation.* Neural Computation 4, 415–447.

Mittal, G.S., Zhang, J., 2000. *Use of Artificial Neural Network to Predict, Temperature, Moisture and Fat in Slab-shaped Foods with Edible Coatings during Deep Fat Frying.* Journal of Food Science 65, 978–983.

Ni, H., Gunasekaran, S., 1998. *Food Quality Prediction with Neural Networks.* Food Technology 52, 60–65.

Ruhil, A.P., Singh, R.R.B., Jain, D.K., Patel, A.A., Patil, G.R., 2009. *Development of An Artificial Neural Network based Model for Shelf-life Prediction of Basundi Mix: An Indian Dairy Product.* In: Proceedings of the 3rd Indian International Conference on Artificial Intelligence – 2007, Pune, India, December 17-19, pp. 1517-1524.

Sinclair, C., 2004. *Dictionary of Food.* A & C Black, London.

Singh, R.R.B., Ruhil, A.P., Jain, D.K., Patel, A.A., Patil, G.R., 2009. Prediction of Sensory Quality of UHT Milk–A Comparison of Kinetic and Neural Network Approaches. Journal of Food Engineering 92, 146–151.

Sofu, A., Ekinci, F.Y., 2007. *Estimation of Storage Time of Yogurt with Artificial Neural Network Modelling.* Journal of Dairy Science 90(7), 3118–3125.

Vallejo, B., Arteaga, G.E., Nakai, S., 1995. *Predicting Milk Shelf Life based on Artificial Neural Networks and Headspace Gas Chromatographic Data.* Journal of Food Science 60, 885–888.

OO

3

Effect of Partially Hydrolyzed Guar Gum on Quality of Cookies

Deepak Mudgil, Sheweta Barak and B.S. Khatkar

Abstract

Fortification of bakery products with dietary fiber is attracting interest of food scientists and technologists. Present study involves the investigation of possibility of using guar gum soluble fiber in cookies. In last few decades, research studies evidenced that partially hydrolyzed guar gum is useful in heart diseases, diabetes and digestive problems. The study was aimed to investigate the effect of guar gum soluble fiber at 1-5% level on physical and sensory properties of cookies. Spread ratio and overall sensory acceptability of cookies first increased upto 2% addition of partially hydrolyzed guar gum, thereafter decreased significantly whereas hardness of cookies increased significantly with guar gum soluble fiber addition. This study revealed that partially hydrolyzed guar gum at 2% level can be successfully incorporated in cookies for the enrichment of soluble dietary fiber without disturbing the physical and sensory properties of cookies.

Keywords*: guar gum; soluble dietary fiber; cookies; spread factor, cookie hardness*

Introduction

In processed food industry, guar gum served as thickening, stabilizing and texturizing agent in many processed food products such as tomato ketchup, dairy products, processed fruit juices, bakery products. Guar gum can perform beneficial physiological functions in human due to its complex carbohydrate known as galactomannan which is made up of galactose and mannose units. It is beneficial in lowering cholesterol level, control diabetes and regulate bowel digestive system in human beings. Galactomannan is water soluble in nature and cannot be digested by our intestinal secretions hence serve as soluble dietary fiber when consumed. Native guar gum cannot be incorporated as such in food products due to its viscous nature which ultimately deteriorates the sensory and technological properties of food products. Hence Partially Hydrolyzed Guar Gum (PHGG) is produced via enzymatic hydrolysis of guar gum. PHGG thus obtained, is a low viscosity water soluble gum with many physiological health benefits. Enzymatic hydrolysis reduces viscosity of guar gum via molecular weight and chain length reduction. According to AACC 2001, dietary fiber is the edible plant portions or analogous carbohydrates that are not digested and absorbed in the human small intestine with complete or partial fermentation in the large intestine. Polysaccharides, oligosaccharides, lignin, and associated plant substances may be included in dietary fiber definition. Physiological functions like laxation, blood cholesterol attenuation and blood glucose attenuation may be performed by dietary fiber (AACC 2001). Soluble fiber enrichment of bakery products is in considerable interest as it could overcome the increasing demand of daily fiber intake.

PHGG is tasteless, odorless and give solution of low viscosity with water upto 10 cps. These unique properties of PHGG made it suitable for soluble fiber enrichment of processed food products. Preparation of dietary fiber fortified food products involves a major problem that negatively affects the product's functional properties (Brennan and Samyue 2004). Various attempts have been made by food scientists to improve nutritional properties of cookies by incorporating legumes, proteins, fiber, etc (Singh et al. 1996; Ellis 1985). The purpose of the present study was to understand the effect of PHGG addition at various levels on spread ratio and hardness and overall sensory acceptability of cookies.

Material and Methods

Materials

Commercial food grade guar gum sample was procured from Hindustan Gums & Chemicals Ltd., Haryana, India. guar gum sample was passed through 200 mesh sieve to obtain a uniform particle size fine powder . Before analysis this uniform particle size guar gum powder was stored under refrigerated conditions. All chemical of AR grade were obtained from Central Drug House, India. Cellulase enzyme (*Aspergillus niger*) was obtained from USB Corporation, USA. Refined wheat flour, sugar, hydrogenated vegetable shortening and salt were procured from local market Hisar, India.

PHGG Preparation

Enzymatic hydrolysis of guar gum was done to produce partially hydrolyzed guar gum. Hydrolysis of guar gum was done using cellulase from *Aspergilus niger* at pH 6 in aqueous solution maintained at 50°C. After enzymatic hydrolysis, low viscosity guar gum aqueous solution was obtained which was freeze dried, ground and passed through 200 mesh sieve to obtain uniform particle size fine powder as that of native guar gum.

Cookie Preparation

Refined wheat flour was replaced with partially hydrolyzed guar gum at levels of 1%, 2%, 3%, 4% and 5%. Control cookies were prepared by white flour. The ingredients in standardized formulation of cookies were 100 g of flour blend, 25 g of milk, 42 g of fat, 60 g of sugar, 4 g of water, 1.2 g of Ammonium Bicarbonate (ABC), 0.5 g of Sodium Bicarbonate (SBC). Ground sugar, fat, water, SBC and ABC were mixed for 2 min in a dough mixer at medium speed to get a uniform creamed mixture and then milk and flour blend were added at mixed for 3 min at medium speed. The dough obtained was sheeted on dough sheeter into a 10 mm thick sheet. It was cut into round shape with cutter and baked at 185°C for 10 min. After baking, cookies were allowed to cool at room temperature for 3 h and then subjected to physical and sensory analyses. Effect of water level from 4% to 8% and baking time 8-12 minutes on cookies supplemented with PHGG from 1% to 5% were studied.

Cookie Spread Factor

Spread factor of the cookies was determined according to methods of AACC (1995). Spread factor of cookies were measured from the ratio of diameter and thickness.

Cookie Texture

Hardness of cookies was measured with Texture Analyzer (model TA-XT2i, Stable Micro systems, U.K.) using blade set with knife. Testing conditions for textural studies were pre-test speed of 1.5 mm/sec, test speed of 2.0 mm/sec, post test speed of 10 mm/sec, distance 10 mm, trigger force of 25 g, and load cell of 50 kg. The peak value of maximum fracture force was considered as hardness when the cookie broken into two major pieces (Chakraborty et al. 2009).

Breaking strength of cookies was represented by this peak force value (g). Mean values in triplicates are reported as fracture force.

Sensory Evaluation

Sensory properties of cookies were evaluated by a sensory panel of 10 semi-trained panel members. Freshly made cookies were used for evaluation of sensory attributes. Panelists were asked to give score to color, flavor, mouth feel and texture. Overall acceptability of cookies was estimated by taking average of all the above sensory parameters. A nine-point hedonic scale (1 = dislike extremely, 2 = dislike very much, 3 = dislike moderately, 4 = dislike slightly, 5 = neither like nor dislike, 6 = like slightly, 7 = like moderately, 8 = like very much and 9 = like extremely) was used by the panelists to evaluate sensory properties of experimental cookies.

Results and Discussion

Spread Ratio

Spread ratio is a physical property which is a measure of cookie quality. Cookies with higher spread are considered better in quality. Spread ratio of cookies showed first increasing and then decreasing trend with PHGG addition level. Upto 2% concentration level, spread ratio increased from 7.02 to 7.56. This increase in spread ratio is due to low gluten formation because of dilution effect. In PHGG supplemented cookies, wheat flour was replaced with PHGG. It diluted the wheat flour concentration which resulted in low gluten formation and higher spread ratio. Above 2% concentration spread ratio decreased from 6.6 to 5.06. This decrease in spread ratio value may be due to the viscosity effect of PHGG. At higher concentration PHGG increased viscosity of cookie dough which resulted decrease in spread ratio. Rajiv *et* al (2011) reported that incorporation of flaxseed in cookies decreased the spread ratio and enhanced the hardness in cookies.

TABLE 3.1

Effect of PHGG Addition on Cookie Qualities

PHGG Level	*Spread Ratio**	*Hardness**	*Overall Acceptability***
0%	7.16	7233.1	8.5
1%	7.02	7362.9	8.7
2%	7.56	9521.1	8.2
3%	6.6	9607.0	7.5
4%	4.95	9720.5	6.6
5%	5.06	11818.2	6.8

*values are mean of three independent measurements.
**values are mean of ten independent measurements

Hardness

Hardness is also a physical property which is a measure of cookie quality. Softer cookies are considered better in quality. Hardness of cookies increased with increase in addition of PHGG level. This increase in hardness is due to the fiber property of PHGG. Addition of fiber increases hardness in cookies (Rajiv *et al* 2011). Dietary fiber components such as water soluble pentoses, pectin, water insoluble hemicelluloses, cellulose and lignin from wheat bran, oat bran, corn bran, navy bean hull and soy hulls significantly reduced cookie spread and tenderness of cookies (Jeltema *et al* 1983).

Overall Sensory Acceptability

A sensory evaluation panel of 10 members judged the appearance, color, aroma, taste, texture, and overall acceptability of the cookies fortified with guar gum fiber. Compared with control samples, overall sensory acceptability of cookies fortified with guar gum fiber decreased with increase in concentration of partially hydrolyzed guar gum. In the current study, desirable results for overall sensory acceptability of cookies were obtained at 3% level of incorporation.

Conclusion

Cookies supplemented with PHGG at 2.0% level resulted in cookies with low tenderness, little low acceptability and higher spread ratio which is significant from the point of view of commercial utilization of partially hydrolyzed guar gum for dietary fiber enrichment of cookies without disturbing the physical and sensory quality but with little improvement in physical characteristics of cookies.

References

AACC. (1995). Approved Methods of the American Association of Cereal Chemists. USA: American Association of Cereal Chemists.

AOAC. (1990). Official Methods of Analysis (15th ed.). Washington: Association of Official Analytical Chemists.

Brennan, C.S., Samyue E. (2004) *Evaluation of Starch Degradation and Textural Characteristics of Dietary Fiber enriched biscuits.* Int J of Food Prop 7(3):647–657.

Chakraborty, S.K., Kumbhar B.K., Chakraborty S., Yadav, P. (2011) *Influence of Processing Parameters on Textural Characteristics and Overall Acceptability of Millet enriched Biscuits using response Surface Methodology.* J Food Sci Technol 48(2):167-174.

Chakraborty, S,K,, Singh, D,S,, Kumbhar, B,K,, Singh, D. (2009) *Process Parameter Optimization for Textural Properties of Ready-to-eat extruded Snack Food Form Millet and Pulse-brokens Blends.* J Texture Stud 40(6):710–726.

Design Expert Version 8.0.4.1, 2010. Stat-Ease, Inc., MN, USA

Ellis, P.R. (1985) *Fiber and Food Products.* In: Leeds AR, Avenell A (eds) Dietary Fiber Perspectives: Reviews and Bibliography. John Libbey Company Limited, London, pp 83–105.

Frost, D.J., Adhikari, K., Lewis, D.S. (2011) *Effect of Barley Flour on Physical and Sensory Characteristics of Chocolate Chip Cookies.* J Food Sci Technol. DOI 10.1007/s13197-010-0179-x

Jeltema, M.A., Zabik, M.E., Thiel, L.J. (1983) *Prediction of Cookie Quality from Dietary Fiber Components.* Cereal Chem 60(3):227-230.

Montgomery, D.C. (1984) *Design and Analysis of Experiments.* John Wiley & Sons, Inc, Singapore

Myers, R.H. and Montgomery, D.C. (1995) *Response Surface Methodology: Process and Product Optimization using Designed Experiments.* New York: John Wiley & Sons, Inc.

Rajiv, J., Indrani, D., Prabhasankar, P., Rao, G.V. (2011) *Rheology, Fatty Acid Profile and Storage Characteristics of Cookies as influenced by Flax Seed* (*Linum usitatissimum*). J Food Sci Technol. DOI 10.1007/s13197-011-0307-2

Singh, R., Singh, G., Chauhan, G.S. (1996) *Effects of Incorporation of Defatted Soy Flour on the Quality of Biscuits.* J Food Sci Technol 33(4):355–357.

Weng, W., Liu, W., Lin, W. (2001) *Studies on the Optimum Models of the Dairy Product Kou Woan Lao using response Surface Methodology.* Asian Austral J Anim 14(10): 1470–1476.

OO

Newer Techniques in Processing of Rice Bran Oil and its Blends with a Special Reference to Quantification

H.K. Sharma

Abstract

RBO (Rice Bran Oil) is an excellent cooking and salad oil due to its high smoke point and delicate flavor. The nutritional qualities and health effects of rice bran oil are also established. The content of phospholipids in crude rice bran oil is major problems while processing of the oil in various food applications. Water degumming process is suitable for the removal of Hydratable Phospholipids (HPL) whereas Non-Hydratable Phospholipids (NHPL) can be removed only during acid degumming or enzymatic degumming processes. Efficiency of enzymatic degumming was evaluated and compared with acidic degumming on the basis of phosphorous content, FFA (free fatty acid), color of degummed oil and residual oil content in gums separated. Enzymatic degumming reduces phosphorous content of the oil from 350 to 42 ppm while it was reduced to 86 ppm after acidic degumming. The oil residue, 30.81% in the separated gums was observed after enzymatic degumming which was found lesser than 51.72% oil residues in gums after acidic degumming of oil. Quantification of pure oil in blended oils is another challenging technique for researchers. Pure rice bran oil was quantified in its blends with safflower and sunflower oil on the basis of changes in physicochemical parameters and correlation between them. The study revealed that regression equations based on the oryzanol content, palmitic acid composition, ultrasonic velocity, relative association, acoustic impedance, and iodine value can be used for the quantification of rice bran oil in blended oils. The rice bran oil can easily be quantified in the blended oils based on the oryzanol content by HPLC even at 1% level. The same set of blended oil were exposed to frying procedure and the changes occurring in the rice bran oil and its blend with sunflower oil during repeated frying cycles of dried and moist potato chips were monitored. The oryzanol content and iodine value decreased with the frying cycles. The increase in p-anisidine value was more in rice bran oil as compared to blended oil. The amount of unsaturated fatty acid decreased gradually during repeated deep fat frying cycles in both the oils. The trans fat increased with repeated deep fat frying cycles in both the rice bran and blended oils, when used to fry moistened and dried potato chips. Both the oil samples showed greater formation of trans fatty acids when the moistened potato chips were used during frying

Introduction

Rice Bran Oil (RBO) is an excellent cooking and salad oil due to its nutritional qualities, health effects, high smoke point and delicate flavour. RBO is rich in unsaponifiable fraction (unsap 4.2%) which includes antioxidants and micronutrients, like vitamin E complexes, gamma oryzanol, phytosterols, polyphenols and squalene. The phospholipids present in RBO are predominantly hydratable Phosphatidyl Choline (PC), Phosphatidyl-Inositol (PI) and non-hydratable phospholipids are calcium and magnesium salts of Phosphatidic Acid (PA) and Phosphatidyl Ethanolamine (PE) (Hvolby, 1971). Phospholipids should be removed because of their strong emulsifying action and if they are not removed, the oil will went through undue darkening during deodorization at high temperature (Kim et al., 2002). The high oxidative stability of RBO makes it preferred oil for frying and baking applications (McCaskill and Zhang 1999, Semwal and Arya 2001). Investigations pertaining to the degumming of rice bran oil and processing are scarce and need focus of researchers

Enzymatic Degumming

Rice bran oil differs from other vegetable oils because of its higher Free Fatty Acid (FFA) content along with its unusually high contents of wax, unsaponifiable constituents, polar lipids (including glycolipids), and pigments (Kitts, 1996). The typical composition of crude RBO is 81–84% triacylglycerols (TAG), 2–3% Di-Acyl-Glycerols (DAG), 1–2% Mono-Acyl-Glycerols (MAG), 2–6% Free Fatty Acids (FFA), 3–4% wax, 0.8% glycolipids, 1–2% Phospho-Lipids (PL) and 4% unsaponifiable matters (Hvolby, 1971). The phospholipid content of crude oil should be removed concerning the quality of oil. Traditional degumming processes like water and acid degumming are found successful in removing only hydratable phospholipids where as nonhydratable phospholipids left untreated throughout the degumming process. Enzymatic degumming converts non-hydratable lecithin (gums) to water-soluble lysolecithin, which is separated by centrifugation and hence provide better yield. The process uses relatively small amounts of citric acid and caustic along with a phospholipase (Lecitase® Ultra) with about 2% water (Chakrabarti & Rao, 2004). The free fatty acids are then removed in the deodorization step and can be used as a valuable co-product, or further processed into the value added products, such as biodiesel fuel (Zufarov et al., 2009). Enzymatic method of degumming vegetable oil is famous under the name Lurgi's EnzyMax®process, which makes use of phospholipase (Clausen, 2001). The process reduces phosphorous content of the oil, a measure of residual gums, to below 5 ppm. It generates less waste, both water and soapstock and increases oil yield, reducing environmental impact versus traditional processing.

This process consists of three important steps: adjusting the pH of the oil with buffer, the addition of the enzyme solution and carrying out the enzyme reaction, and separating gum/sludge from the oil. Mishra et al, (2011) evaluated the efficiency of enzymatic degumming over acid degumming of rice bran oil on the basis of phosphorous content, FFA, color of degummed oil and residual oil content in gums separated. This comparative study can be clearly depicted through (Table 4.1). This process of enzymatic degumming reduces phosphorous content of the oil from 350 to 42 ppm, whereas the phosphorous content after acidic degumming was reduced up to 86 ppm. Recovery of oil was found more after enzymatic degumming. The oil residue, 30.81% was observed in the enzymatically separated gum which was found lesser than 51.72% oil residues in gums after acidic degumming of oil (Mishra et al., 2011).

TABLE 4.1

Changes in Physicochemical Parameters of Enzymatic and Acidic degummed Rice Bran Oil

Composition	*Enzymatic Degumming*	*Acidic Degumming*
P content	Reduced from 350 to 42 ppm	Reduced from 350 to 86 ppm
FFA	Reduced from 16.09 to 16.07%	Increased from 16.29 to 16.60
Color	Reduced from 56.0 to 50.5 unit	Reduced from 56.0 to 38.0
Residual oil in gum	30.81%	51.72%

p-phohphorous, FFA-Free Fatty Acid, ppm-parts per million

Formation of Trans Fats during Thermal Processing of Rice Bran Oil

During the frying process, the oil is continuously or repeatedly used at elevated temperatures in the presence of air. Thus, both thermal and oxidative decomposition of the oil may take place (Chang *et al.,* 1978, Pokorny, 1998, Nawar, 1998). Advanced oxidation, due to high frying results in the formation of trans fatty acids (Kiritsakis *et al.,* 1989; Ovesen *et al.,* 1998).

Sharma et al., (2011) blended rice bran oil with sunflower and safflower oil and subjected these blends for 6 consecutive frying cycles. These blends were then analyzed for various physichochemical parameters like color, Free Fatty Acid (FFA), Peroxide Value (PV), Iodine Value (IV), Anisidine Value (PAV), Saponification Value (SV) and Trans Fats.

TABLE 4.2

Behavior of Rice Bran and Blended Oil
(RBO: SUNF – 60:40) after Frying the Potato Chips having Moisture 0.05% (wb)

Parameter	*Sample*	*Control*	*After 1st frying*	*After 2nd frying*	*After 3rd frying*	*After 4th frying*	*After 5th frying*	*After 6th frying*
FFA	RBO	0.08^{a}	0.09^{acf}	0.09^{acfh}	0.1^{acfhk}	0.11^{acfhkl}	$0.12^{bcfhklm}$	$0.13^{bdgjklm}$
	Blend	0.07^{d}	0.09^{dg}	0.09^{dgi}	0.1^{dgik}	0.1^{dgikl}	0.11^{fgiklm}	0.13^{fhjklm}
Colour	RBO	10^{a}	10.5^{ac}	10.5^{ace}	11.5^{aceg}	12^{acegh}	13^{bdfghj}	13.5^{bdfghj}
	Blend	9.0^{a}	9.0^{af}	9.5^{afh}	10.0^{afhj}	10.5^{afhjk}	10.5^{afhjkl}	11.5^{dgijkl}
PV	RBO	0.59^{a}	1.02^{af}	1.58^{afh}	2.04^{afhk}	2.57^{bfhkl}	2.99^{bghklp}	3.61^{bgjklp}
	Blend	1.49^{a}	1.89^{ae}	2.43^{aeg}	2.85^{aegk}	3.07^{aegkl}	3.66^{dfgklt}	4.01^{dfgklt}
IV	RBO	99.88^{a}	98.19^{ad}	97.12^{adf}	95.31^{adfh}	93.11^{cdfhk}	92.79^{cdfhkt}	90.21^{ceghkt}
	Blend	105.60^{a}	103.06^{ac}	100.65^{acf}	99.15^{acfg}	98.09^{acfgh}	96.57^{befghj}	94.91^{befghj}
p-anisidine	RBO	46.02^{a}	46.21^{ac}	46.67^{acf}	48.33^{acfg}	51.04^{bcfgh}	53.94^{befghk}	56.45^{befghk}
	Blend	39.11^{a}	39.85^{ac}	41.03^{acf}	43.47^{acfh}	46.43^{acfhk}	50.01^{bcfhkl}	53.02^{beghkl}
Trans Fat (%)	RBO	1.27^{a}	1.51^{ac}	1.69^{acf}	1.8^{acfg}	1.99^{bcfgh}	2.18^{bcfghj}	2.37^{befghj}
	Blend	1.15^{a}	1.25^{ad}	1.32^{adf}	1.4^{adfh}	1.53^{adfhk}	1.66^{befhkl}	1.80^{begjkl}

*Means within the same rows sharing a common small letter are not significantly different at $P<0.05$.FFA- free fatty acid, colour, PV-peroxide value, IV-iodine value, p-anisidine value, OC-Oryzanol content, oil uptake. Results are average of three individual experiments.

TABLE 4.3

Behavior of Rice Bran and Blended Oil (rbo: sunf – 60:40) after Frying the Potato Chips having Moisture 64.77% (wb)

Parameter	*Sample*	*Control*	*After 1st frying*	*After 2nd frying*	*After 3rd frying*	*After 4th frying*	*After 5th frying*	*After 6th frying*
FFA	RBO	0.08^{a}	0.09^{ac}	0.1^{ace}	0.1^{acef}	0.12^{bcefg}	0.13^{bcefgh}	0.14^{bdefgh}
	Blend	0.07^{a}	0.09^{af}	0.1^{afh}	0.13^{afhk}	0.14^{bfhkl}	0.16^{bghklt}	0.17^{bgjklt}
Colour	RBO	10^{a}	10.5^{ac}	12^{acf}	12.5^{acfg}	13^{bcfgk}	13^{bcfgkl}	14.5^{bdfgkl}
	Blend	9^{a}	10^{ae}	10.5^{aeg}	11^{aegh}	11.5^{aeghj}	12.5^{ceghjk}	13^{cfghjk}
PV	RBO	0.59^{a}	1.86^{ac}	3.39^{acf}	4.51^{acfh}	6.07^{bcfhk}	7.61^{beghkl}	8.56^{begjkl}
	Blend	1.49^{a}	2.32^{af}	3.17^{afk}	3.92^{afkl}	4.68^{afklm}	5.26^{dfklmt}	6.59^{dgklmt}
IV	RBO	99.88^{a}	98.47^{aq}	96.66^{aqs}	95.91^{aqsy}	92.52^{pqsyl}	91.81^{pqsylt}	89.02^{prtylt}
	Blend	105.60^{b}	103.43^{bd}	101.09^{bdf}	98.21^{bdfk}	96.53^{bdfkl}	93.03^{cefklq}	90.79^{cegklq}
p-anisidine	RBO	46.02^{a}	53.21^{af}	59.43^{afh}	63.96^{efhj}	68.51^{efhjk}	73.69^{efhjkl}	78.64^{eghjkl}
	Blend	39.11^{a}	44.44^{af}	48.58^{afi}	51.47^{afik}	59.88^{dfikl}	65.36^{dfiklm}	70.71^{dgjklm}
Trans Fat (%)	RBO	1.27^{a}	1.49^{ac}	1.69^{acg}	1.96^{acgj}	2.21^{acgjk}	2.64^{bfgjkl}	2.91^{bfhjkl}
	Blend	1.15^{a}	1.36^{ac}	1.58^{acg}	1.65^{acgl}	1.86^{acglm}	2.23^{bfglmp}	2.48^{bfhlmp}

*Means within the same rows sharing a common small letter are not significantly different at $P<0.05$.FFA- free fatty acid, colour, PV-peroxide value, IV-iodine value, p-anisidine value, OC-Oryzanol content, oil uptake. Results are average of three individual experiments.

The changes in these physichochemical parameter after each frying of potato chips of varying moisture content (0.05% and 64.77) is tabulated as table 4.2 & 4.3. Both the blended oil samples showed greater formation of trans fatty acids when the moistened potato chips were used frying (Table 4.4 & 4.5). These data clearly revealed the rapid degradation of the quality of oils, when used to fry the products consisting of higher moisture. The blended oil samples were found better as the samples were acceptable after even sixth repeated deep fat frying, when used to fry dried potato chips however the significant differences ($P < 0.05$) in the total trans fats was observed after 5th repeated deep fat frying cycles.

Newer Techniques for the Quantification of Rice Bran Oil in Blended Oils

The various advantages of blended oils and their population in market have also raised the need of rapid and simple technique to quantify proportion of specific oil in a blend. Hence it is essential to have rapid and simple procedure to detect the proportion of specific oil in a blend. Various methods have been reported for detection of oil.

1. Simple Techniques

The simple technique includes the physicochemical (Iodine value, peroxide value, saponification value, free fatty acid composition, refractive index and specific gravity) analysis of pure and blended oil. The changes in percent concentration of oil in the blend to be detected and the changes in these physicochemical parameters are correlated. The physicochemical parameters found with significant changes are then selected as parameters to quantify the oil in its blends.

Sharma et al., 2011 selected Iodine value for quantifying rice bran oil in its blends with safflower oil. The changes in IV on increasing the proportion of RBO in blend is provided in Table 4.4.

2. Chromatographic Techniques

Chromatographic techniques for quantification of single oil in its blends includes:

Gas Chromatography: Fatty acid profile of pure oils and their blends is determined by this method. The changes in proportion of pure oil in its blends and the changes in fatty acid profile are correlated. The fatty acid reflecting significant changes along with the changes in proportion of pure oil is selected as the parameter for quantification of that oil. Sharma et al., 2011 quantified rice bran oil in its blends by estimation of fatty acids content through gas chromatography. Increase in proportion of PRBO (pure rice bran oil) in its blends resulted an increase in palmitic acid content of blended oils. The data in Table 4 reflected that there was significant difference in estimated values of palmitic acid content of PRBO, SnF (sunflower oil) and SAF (safflower oil) thus palmitic acid content of blended oils was considered as an independent parameter for quantification of PRBO.

High Pressure Liquid Chromatography: The method can be used for quantitative detection of major components like minerals, antioxidants, vitamins etc, in samples. The oryzanol present in rice bran is reported to have functions similar to vitamin E in promoting growth, facilitating capillary growth in the skin, and improving blood circulation along with stimulating hormonal secretion. Sharma et al., (2011) estimated the oryzanol content in rice bran oil through HPLC and spectrophotometry method. The data in Table 4 indicated the difference in oryzanol content of oils and their blends when were subjected to Spectrophotometry and HPLC detection. The HPLC and spectrophotometric estimation of oryzanol content can quantify even at a level of 1 and 4% of PRBO in blended oils respectively. The oryzanol was found higher in all the samples when estimated by HPLC which may be due to higher accuracy and sensitivity of the method as compared to spectrophotometric method.

Application of Ultrasonic Velocity

Ultrasonic velocity has been regarded as an important tool for the evaluation of several physical properties of oils and fats.

Ultrasonic velocities, relative association, acoustic impedance and compressibility were calculated at 1 and 2 MHz frequencies (Sharma et al., 2011)

The isotropic compressibility, relative association, acoustic impedance and adiabatic compressibility were calculated by using equation 1, 2, 3 & 4 respectively (Mehra and Israni, 1999).

$$Ks = u^{-2}\rho^{-1} - (1),$$

$$RA = (\rho/\rho^0)\ (u^0/u)^{1/3} - (2),$$

$$Z = u\ \rho - (3),$$

$$\beta = 1/\ (u^0)^2\ \rho^0 - (4)$$

Ks isotropic compressibility: ρ^0 densities of control sample: U^0 ultrasonic velocity of control sample: RA relative association: Z acoustic impedance: β adiabatic compressibility: ρ density of sample: u ultrasonic velocity of sample.

All these parameters reflected a significant distinguished change at frequency of 2MHz in pure oils and their blends (Table 4). In this context ultrasonic velocities, acoustic impedance and relative association at 2MHz were considered as independent quantifying parameters in estimation of PRBO content.

TABLE 4.4

Physicochemical Parameters of Pure Rice Bran Oil, Sunflower Oil, Safflower Oil and their Blends

Composition	*IV*	*PA*	*USV*	*Ra*	*Z*	*OC by HPLC*	*OC by SPM*
Pure RBO	101.55±0.08	19.09±0.02	690	0.0996	746.33	13897.39±2.76	13970.9±2.43
Pure Sunflower	113.16±0.02	7.97±0.04	731	0.1039	661.68	ND	ND
SnF+ PRBO (99+1)	113.16±0.10	8.69±0.07	728	0.1039	661.68	169.87	ND
SnF +PRBO (98+2)	112.73±0.04	8.78±0.05	727	0.1038	662.60	356.11	ND
SnF +PRBO (97+3)	112.06±0.04	8.27±0.02	725	0.1038	662.60	601.27	ND
SnF +PRBO (96+4)	111.32±0.05	9.02±0.08	724	0.1038	663.52	996.78±1.43	644.17±1.61
SnF+ PRBO (95+5)	111.05±0.05	9.51±0.04	723	0.1037	664.44	1012.31±1.64	907.64±1.60
SnF +PRBO (90+10)	109.68±0.05	9.95±0.07	721	0.1035	669.03	1561.92±1.88	1463.41±1.76
SnF+ PRBO (80+20)	107.91±0.04	10.41±0.03	720	0.1030	672.56	2602.71±1.43	2519.28±1.27
Pure Safflower oil	137.63±0.03	5.78±0.02	798	0.1012	721.14	NIL	NIL
SAF+ PRBO (99+1)	137.47±0.05	5.79±0.07	795	0.1012	722.06	117.85±2.21	NIL
SAF +PRBO (98+2)	137.39±0.08	6.12±0.03	794	0.1011	722.98	298.72±1.86	NIL
SAF +PRBO (97+3)	137.32±0.05	6.25±0.03	792	0.1011	723.91	408.45±1.14	NIL
SAF +PRBO (96+4)	137.15±0.06	6.34±0.02	790	0.1010	726.67	569.56±1.46	412.18±1.62
SAF +PRBO (95+5)	137.08±0.06	6.66±0.08	789	0.1009	729.43	658.73±1.31	675.93±1.28
SAF+ PRBO (90+10)	136.93±0.04	6.89±0.07	786	0.1007	730.59	1312.59±1.39	1354.11±1.56
SAF +PRBO (80+20)	136.63±0.08	7.52±0.02	783	0.1004	733.26	2729.66±1.29	2701.82±1.35

PA- Palmitic Acid, IV- Iodine Value, USV- Ultra Sonic Velocity, Ra- Relative Association, Z- Impedence, OC- Oryzanol Content, HPLC- High Pressure Liquid Chromatography, SPM- Spectrophotometry

TABLE 4.5

Regression Equation and Correlation Coefficient of Blended Oils

Constituents	*X- axis*	*Y- axis*	*Equation* $Y = bX + C$	R^2
PRBO and SnF blends	% PRBO in blended oil	Oryzanol content	Y = 122.151 X + 257.742 *	0.959
PRBO and SnF blends	% PRBO in blended oil	Palmitic acid (C_{16}) content	Y= 0.0952 X + 8.660 **	0.767
PRBO and SnF blends	% PRBO in blended oil	Ultrasonic velocity at 2 MZ	Y = -0.384X + 726.468*	0.753
PRBO and SnF blends	% PRBO in blended oil	Relative association at 2 MZ	Y= -0.0000461X + 0.104*	0.992
PRBO and SnF blends	% PRBO in blended oil	Acoustic impedence (Z) at 2 MZ	Y = 0.596 X + 661.377*	0.964
PRBO and SnF blends	% PRBO in blended oil	Iodine Value	Y = -0.265X + 112.836*	0.933
PRBO and SAF blends	% PRBO in blended oil	Oryzanol content	Y= 135.657 X + 1.284*	0.999
PRBO and SAF blends	% PRBO in blended oil	Palmitic acid (C_{16}) content	Y = 0.0906 X + 5.872**	0.872
PRBO and SAF blends	% PRBO in blended oil	Ultrasonic velocity at 2 MZ	Y = -0.601 X + 793.718*	0.864
PRBO and SAF blends	% PRBO in blended oil	Relative association at 2 MZ	Y = - 0.0000419 X + 0.10*	0.964
PRBO and SAF blends	% PRBO in blended oil	Acoustic impedence (Z) at 2 MZ	Y = 0.609 X + 722.789*	0.797

**Regression equation is applicable at or above 20% level of PRBO in blended oil.

* Regression equation is applicable at or above 1% level of PRBO in blend

SnF Sunflower oil: SAF Safflower oil: PRBO Physically refined rice bran oil

Statistical Method of Quantification

To explore the possibility of using all the mentioned parameters having significant changes in values, were used for quantification of pure rice bran oil in the blended oils (PRBO: SnF & PRBO: SFA) under study by subjecting the data to linear regression (Sharma et al., 2011). Regression analysis deals with situations where variation of one variable is dependent on the variation of a second variable. Regression may be positive or negative depending upon the orientation of line ($Y = bX + C$) with respect to the axis. The ultrasonic velocity, relative association and acoustic impedance at 2 MHz, oryzanol content, palmitic acid and iodine value were selected and were plotted as dependent parameters on X-axis and percentage proportion of PRBO in the blends was plotted as a independent parameter on Y-axis. The resulting data gave correlation coefficient (R2) which is provided in Table 4.5. The data revealed that the quality parameters, except palmitic acid content for blend containing PRBO and SnF, can be considered for quantification of PRBO by using the corresponding equation (Table 4.5).

Conclusion

Crude oil, extracted from oil seeds is generally refined to remove impurities that can impact the oil's stability, color, and flavor. Traditional degumming processes, including water degumming, super degumming, TDP (Total Degumming Process) degumming, acid treatment and other ones, cannot guarantee the achievement of low phosphorus contents required for physical refining. Enzymatic degumming of vegetable oil reduces phosphorous content of the oil, a measure of residual gums, to below 5 ppm and it also reduces the losses of oil in the gums. The frying process of rice bran oil and its blends should also be monitored in context of trans fats formation after each frying cycles. The study has also revealed the newer techniques of quantifying rice bran oil in its blended oils. The possibility of using statistical techniques like regression linear equation was explored to quantify rice bran oil in its blends. However, study pertaining to the quantification of individual oil by implementing ultrasonic velocities, acoustic impedence and relative association is further required to be explored in depth.

References

Chakrabarti, P.P., & Rao, B.V. S.K. (2004). Process for the pre-treatment of Vegetable Oils for Physical Refining. USP 005,399.

Chang, S.S., Peterson, J.R.Y, Ho, C. (1978). *Chemical reactions involved in the deep-fat frying of foods*—J Am Oil Chem. Soc, 55, 728-727.

Clausen, K. (2001). *Enzymatic Oil-Degumming by a Novel Microbial Phospholipase*, European Journal of Lipid Science and Technology, 103, 333–340.

Hvolby, A. (1971). *Removal of Nonhydratable Phospholipids from Soybean Oil.* Journal of American Oil Chemist Society, 48, 503–509.

Kim, I.C., Kim, J.H., Lee, K.H., Tak, T.M. (2002). *Phospholipids Separation (degumming) from Crude Vegetable Oil by Polyimide Ultrafiltration Membrane*. Journal of Membrane Science (205): 113-123

Kiritsakis, A., Aspris, P.Y., Markakis, P. (1989). *Trans Isomerization of Certain Vegetable Oils during Frying*, Flavors and Off-Flavors. G. Charalambous, (Ed). Proceeding of the 6 International Flavor Conf. Rethymnon, Crete, Greece.

Kitts, D. (1996). *Toxicity and Safety of Fats and Oils*, Bailey's Industrial Oil and Fat Product, New York John Wiley & Sons, Vol. 1 pp. 215–280.

McCaskill, D.R., & Zhang, F. (1999). *Use of Rice Bran Oil in Foods*. Food Technology Journal article, Chicago.53 (2): 50-53.

Mehra. R. and Israni, R. (1999). *Thermodynamic Parameters of Multicomponent Liquid Mixtures and their Variation with Temperature from Ultrasonic and Density Measurements*. International Conference and Exhibition on Ultrasonics (ICEU). New delhi, december 2-4, 1999.

Mishra, R., Sharma, H.K., Sarkar B.C. and Sengar, G. *Enzymatic Degumming of Crude Rice Bran Oil* (Chapter 7), In Bio processing of Foods, Asiatech publications. Pp 99-106

Ovesen, L., Leth, T.Y., Hansen, K. (1998). *Fatty Acid Composition and Contents of Trans Monounsaturated Fatty Acids in Frying Fats and in Margarines and Shortening marketed in Denmark*, J. Am. Oil Chem. Soc, 75, 1079-1083.

Pokorny, J. (1998). *Substrate Influence on the Frying Process*, Grasas y Aceites, 49, 265-270.

Semwal, A.D., & Arya, S.S. (2001). *Studies on the Stability of Some Edible Oils and their blends during Storage*. Journal of Food Science and Technology, India. 38(5): 515-518.

Sharma, H.K., Mishra, R. and Sengar, G. (2011). *Quantification of Rice Bran Oil in Blended Oils*. Grasas y Aceites, In press.

Zufarov, O., Schmidt, S., Sekretar, S., & Cvengros, J. (2009). *Ethanolamines used for Degumming of Rapeseed and Sunflower Oils as Diesel Fuels*. European Journal of Lipid Science and Technology, 111 (10), 985–992.

OO

Measurement of Required Forces to Break the Closed Shell Pistachio Nut

L. Ghorbani and M. Shamsi

Abstract

This study has been conducted to measure the required forces for breaking the closed shell pistachios to reach the kernel. 180 pistachio samples were loaded under compressive pressure loading from three directions: longitudinal axis (x), transverse axis (y), thickness axis (z). The average moisture content of the pistachios under test was 13.57% with a standard deviation of 4.24%. The results showed that the highest rupture force is along z- axis with a maximum value of 741 N. The lowest rupture force is along the x-axis with the maximum value of 168 N. The maximum rupture forces along the x, y and z axis are respectively 700, 631 and 741. Minimum rupture forces along the x, y and z axis are respectively 168, 208 and 204 N. Average absorbed energy in the x, y and z direction was respectively 1.9, 2.032 and 2.78 J. Absorbed Energy is maximum in z direction.

Keywords: *Processing, Splitting, Shelling, pistachio, rupture force.*

Introduction

Pistachio is one of the main agricultural products produced in Iran, particularly in Rafsanjan city and South Khorasan province. Iran produced about 192,269 Mt of Pistachio nuts in 2008 and the total of revenue of pistachio nuts was approximately 635 US$ (FAOSTAT, 2008), So pistachio is one of the valuable agricultural products that have a great effect on the country economy (Razavi et al., 2007). In some years more than 40% of the total yield of a pistachio orchard may stay closed shell. Splitting and shelling machines must be used to break the shell to reach the kernel. Mechanical properties such as rupture force, deformation and rupture energy to break the nut's shell are practical information required for many purposes including the design of splitting and shelling machines. Rupture force represent the minimum force required to break the nut's hard shell. This paper explains the method of measurement of the required forces and energies to break the closed shell pistachio nut to get the kernel. Nazari et al. (2009) determined rupture force, deformation and rupture energy of five varieties of Iranian naturally open (split) shell pistachio nuts and kernels. Samples were forced in three directions under compressive loading according to the directions shown in figure 5.1.

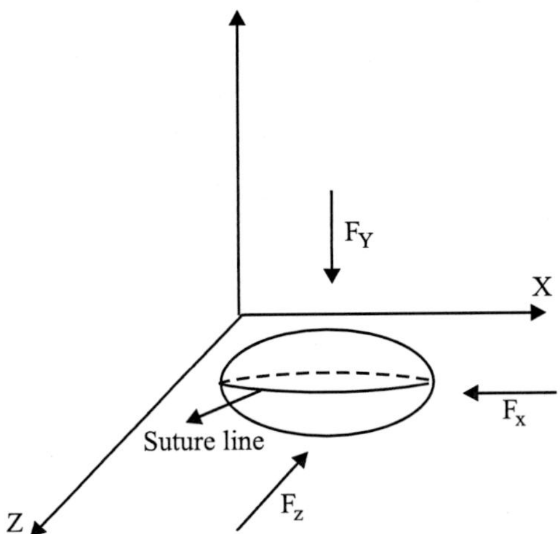

Figure 5.1: *Representation of the Three Axes for Pistachio Nuts Forcing System: Longitudinal Axis (x), A vector which connects the base of the nut to its apex, Transverse or Width Axis (y) which is Perpendicular to the Suture Plane (x-z), Thickness Axis (z) which is in the Suture Plane.*

Longitudinal axis (x) connecting the base of the nut to its apex, transverse or width axis (y) which is perpendicular to the suture plane (x-z), thickness axis (z) which is in the suture plane. Their results showed that the maximum rupture force, deformation and rupture energy values for all varieties were obtained for pistachio nuts loaded along the x-axis. Compression along the z-axis required less compressive force among three compression axes to extract the kernels. Koyuncu et al. (2004) investigate rupture force of walnut before initial rupturing at different compression positions (length, width and thickness).

Experimental results from compression tests indicated that cracking nuts at the length position required less force and yielded the best kernel extraction quality. Altuntas and Erkol (2009) investigate the mechanical properties of walnut. The highest rupture force and rupture power were obtained for walnut cultivars loaded along the width. The lower rupture force and rupture power were obtained with load thickness axis. Aktas et al. (2007) studied the mechanical properties of almond. The maximum force required to initiate pit rupture was found as maximum 554.3 N at z-axis (thickness) and minimum 126.9 N at x-axis (length direction). Fathollahzadeh et al. (2010) analyzed the effects of compression axis on the amount of force needed to crack apricot pits and kernels. Their study also revealed that cracking an apricot pit requires higher rupture force and energy when compressed along its length (x-axis)and that cracking an apricot kernel requires higher rupture force when compressed along its thickness (z-axis).

Results showed that for the proper extraction of kernels, apricot pits should be compressed along its width. Vursavu° and Özgüven (2004) determined the mechanical properties of apricot pit. Results showed that the highest rupture force was obtained for apricot pit loaded along the X-axis (length direction). Güner et al. (2003) placed different varieties of hazelnuts from three directions x (longitudinal), y (transverse or width), z (thickness) under loading. Highest rupture force obtained in the long z-axis (thickness). Kilickan and Güner (2007) determined mechanical properties of olive fruit and kernel. Highest rupture force obtained along the x-axis (length direction).

Materials and Methods

In this study, required forces to fracture closed shell pistachio nuts were studied. For compression tests 180 numbers of different varieties of pistachios were randomly prepared and compressed with instron machine (model STCS 500 Kg C3). Measurement's accuracy of this device was 1N in force and 0.01 mm in deformation. The initial moisture content of each pistachio was determined using oven method at 103 ± 2°C until a constant weight was reached (Nazari et al., 2009). Average moisture content and standard deviation of samples was 13.57% (d.b) and 4.24%, respectively. Experiments were performed on three directions: longitudinal axis (*x*), it is a vector which connects the base of the nut to its apex, transverse or width axis (*y*) which is perpendicular to the suture plane (*x*–*z*), thickness axis (*z*) which is in the suture plane (Figure 5.1). For each axis 60 numbers of pistachios were placed under compressive loading. Loading velocity of machine was set to 5 mm/min. charts of force-Time and force-displacement obtained by Tensa software for each pistachio. In each chart rupture force was determined from force-displacement curve. After fracture, loading was stopped. Absorbed energy by the sample was calculated. It is the area under the force–displacement curve from the beginning up to the breaking point.

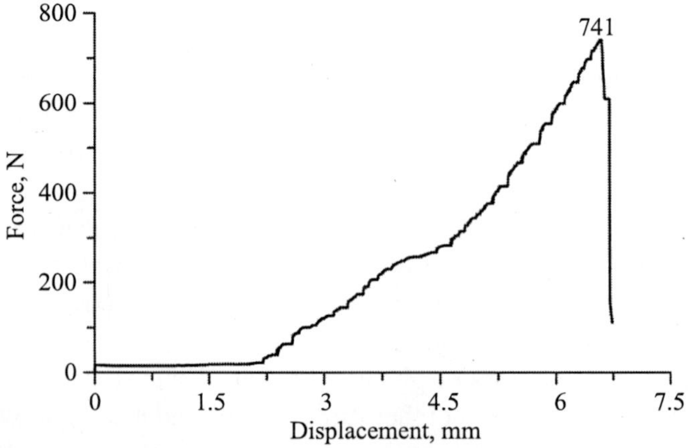

Figure 5.2: *Force-displacement Diagram of Pistachio Nuts under Compressive Force along the x-axis.*

Results and Discussion

A sample of 180 force-displacement Curves of pistachios tested for this study is shown in Figure 5.2.

Measured forces for 10 samples of pistachios in three directions are shown in Table 5.1.

TABLE 5.1

Rupture Forces of 10 Pistachio Samples along the *x*, *y* and *z*-axis

Number	*Fx*	*Fy*	*Fz*
1	461	402	305
2	444	479	250
3	643	248	204
4	368	490	556
5	277	323	282
6	365	366	322
7	340	460	367
8	387	460	335
9	322	577	598
10	256	521	638

Mean and standard deviation of required forces to break the closed shell pistachios in three directions *x*, *y*, *z* are shown in Table 5.2.

TABLE 5.2

Mean and Standard Deviation of Rupture Forces along the *x*, *y* and *z*-axis for 180 Test Samples.

	Fx	*Fy*	*Fz*
Mean	373.5	448.7	481.3
Standard deviation	100.6	96.6	138.5

The maximum value of force in *x*, *y* and *z* axis are respectively 700, 631 and 741 N. The minimum values in *x*, *y*, *z* axis are respectively 168, 208 and 204 N. The results showed that the maximum force value is along the *z*-axis and is equal to 741N. The minimum force value is along the *x*-axis and is equal to 168 N. The tests of Nazari et al, (2009) showed different results. They showed that the maximum rupture force is along the *x*-axis and the minimum rupture force is along the *z*-axis for the nut. The reason is that they have tested the split shell pistachios which had slots along the *z*-axis, so less force is required to break the shell.

The statistical analysis shows significant differences among the rupture forces in three directions at the level of 5%. Three of the Frequency distribution graphs of rupture forces in the three positions are shown in figure 5.3.

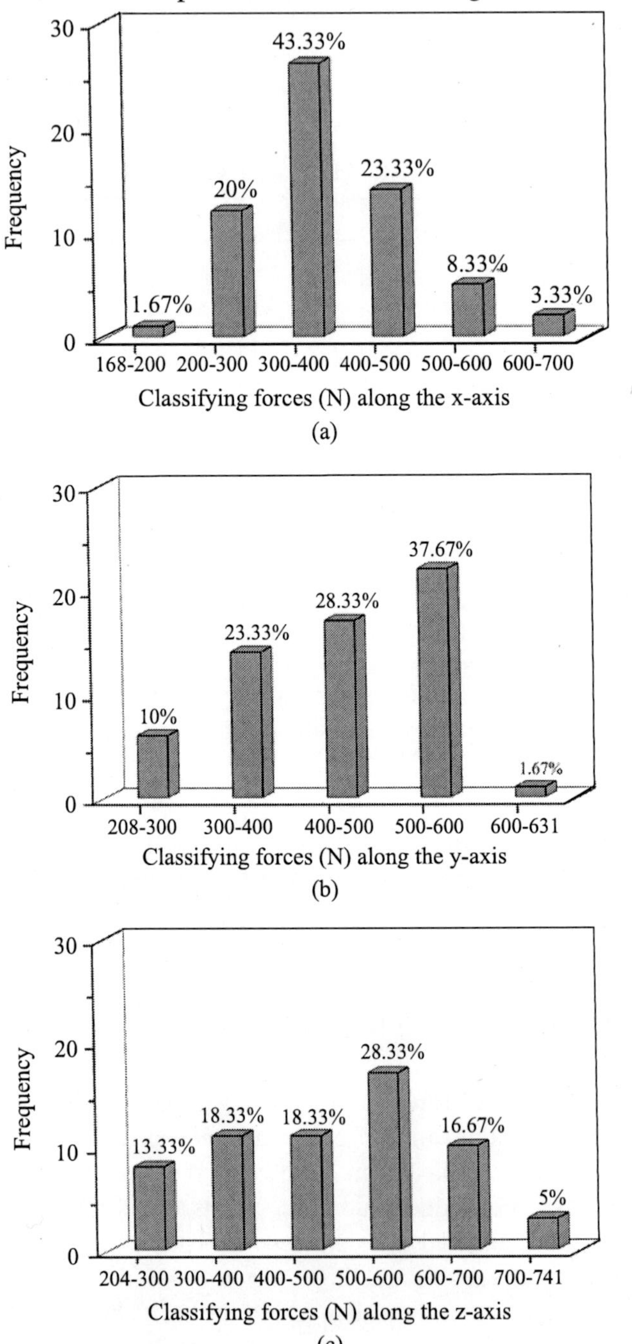

Figure 5.3: *Frequency Diagrams of Forces (a) x, (b) y, (c) z the directions.*

As it can be seen from these figures the maximum and minimum force frequency along the x-axis are 43.3% and 1.67% respectively in the range of 300-400 N and 168-200 N, and the maximum and minimum force frequency along the z-axis are 28.3% and 5% respectively in the range of 500-600 N and 700-741 N, and. the maximum and minimum force frequency in direction y are 36.67% and 1.67% respectively in the range of 500-600 N and 600-631 N. As it can be seen in both of *y* and *z* directions the maximum force frequency are in the range of 500-600N. Exerted force along the z-axis has more uniform distribution relative to *y* and *z* directions. Average absorbed energy in the *x*, *y* and *z* direction was respectively 1.90, 2.03 and 2.78 J. Absorbed Energy is maximum in *z* direction.

Conclusion

Measurement of the required forces and energies to shell the closed shell pistachio is necessary for design of pistachio processing machines including shelling and splitting machines. The compressive loading was applied to 180 samples in this study. The maximum force to crack the shell was 741 N and is in the *z* direction. The maximum force in *x* and *y* direction are respectively 700 and 631 N. Theaverage absorbed energy in the *x*, *y* and *z* direction was respectively 1.90, 2.03 and 2.78 J.

References

Aktas, T., Polat, R., & Atay, U. (2007). *Comparison of Mechanical Properties of Some Selected Almond Cultivars with Hard and Soft Shell under Compression Loading*. Journal of Food Process Engineering, 30, 773–789.

Altuntas E. & Erkol M. (2009). *The Effects of Moisture Content, Compression Speeds, and Axes on Mechanical Properties of Walnut Cultivars*. Food Bioprocess Technol, DOI 10.1007/s11947-009-0283-y.

FAO (2008). Statistical database.http://faostat.fao.org.

Fathollahzadeh, H., Tabatabaie, H., & Mobli, H. (2010). *Effective Conditions for Extracting Higher Quality Kernels from the Sonnatisalams Apricot*. International Journal of Food Engineering, Vol. 6: Iss. 1, Article 1.

Güner, M., Dursun, E., & Dursun, Ý.G. (2003). *Mechanical Behavior of Hazelnut under Compression Loading*. Biosystems Engineering,85 (4), 485–491.

Kibar H. & O¨ztu¨rk T. (2009). *The Effect of Moisture Content on the Physico-Mechanical Properties of Some Hazelnut Varieties*. Journal of Stored Products Research, 45, 14-18.

Kilickan, A., & Güner, M. (2007). *Physical Properties and Mechanical Behavior of Olive Fruits*. Journal of Food Engineering, 87, 222–228.

Koyuncu M.A., Ekinci K., & Savran E. (2004). *Cracking Characteristics of Walnut*. Biosystems Engineering, 87(3): 305-311.

Nazari M., Mohtasebi S.S., Tabatabaeefar A., Jafari A., & Fadaei H. (2009). *Mechanical Behavior of Pistachio Nut and its Kernel under Compression Loading*. Journal of Food Engineering, 95, 499-504.

Razavi M.A., Amini M., Rafe A. & Emadzadeh, B. (2007). *The Physical Properties of Pistachio Nut and its Kernel as a Function of Moisture Content and Variety.* Part III: Frictional properties. Journal of Food Engineering, 81:226-235.

Vursavu° K., Özgüven F., 2004. *Mechanical Behavior of Apricot Pit under Compression Loading. Journal of Food Engineering, 65, 255–261.*

OO

6

Nutritional and Health Benefits of Soybeans and Micro-Enterprise Opportunities

Dr. Suresh Itapu

Introduction

Soybean is one of the nature's wonderful nutritional gifts. Soybean (Glycine max) is a leguminous plant. The Chinese have been cultivating and consuming soybeans for over 4000 years. Soy products have been around in India for almost 30 years. It is one of the very few plants that provide a high quality protein with minimum saturated fat. Soybean helps people feel better and live longer with an enhanced quality of life. India is the fifth largest producer of soybeans in the world. A considerable amount of soybeans in exported. Because of the nutritional and health benefits soybeans are endowed with our effort is to use much more of this food for our selves. In India protein and energy malnutrition is rampant and we are also a predominantly vegetarian country. This makes it even more necessary for us to use soybeans much more regularly in our diet. We realize that adding soy products to our traditional recipes is the best way to go about it and therefore our efforts and activities are focused in this direction.

Nutritional Benefits

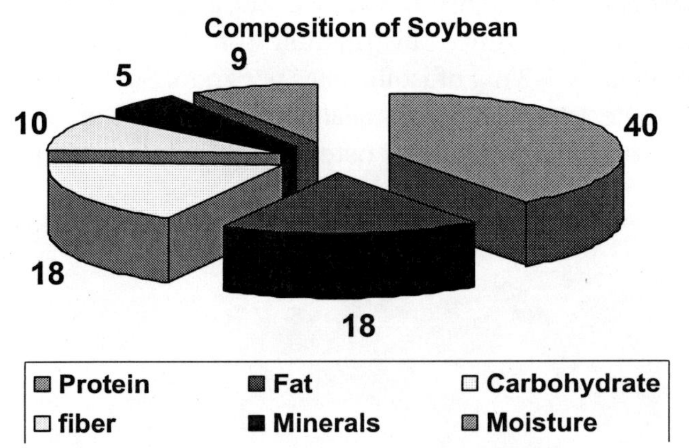

Soybeans contain all the three macronutrients required for good nutrition, as well as fiber, vitamins, minerals. Soybean protein provides all the essential amino acids in the amounts needed for human health. Almost 40 per cent of the calories from soybeans are derived from protein, making soybeans higher in protein than any other legumes and many animal products. Protein in just 250 grams of soybean is equivalent to protein in 3 liters of milk or 1 kg of meat or twenty-four eggs. But it is the quality of soy protein that is most remarkable. Health professionals consider soy protein as superior protein compared to a lot of vegetable proteins and at par with egg and milk protein. The amino acid pattern of soy protein is virtually equivalent in quality to that of milk and egg protein. The 1990's FAO/ WHO protein evaluation committee put soy protein at par with egg and milk protein and ahead of beef protein.

Unlike many other good sources of protein, soybean not only has higher percentage of oil but also quality fatty acid profile. It has low saturated fat content with high amount of poly-unsaturated fat and is a readily available source of essential fatty acids. Soybean oil is also rich in omega-3 and 6 fatty acids similar to those found in fish oils and is cholesterol-free.

Soybean has more than two times the amount of most of the minerals, especially calcium, iron, phosphorus and zinc, than any other legume and very low sodium content. Soybean is a good source of B complex vitamins. Soybeans, especially the outer hull, are an excellent source of dietary fiber. When soybeans are processed, the hull is removed and then processed further to create a fiber additive for breads, cereals and snacks.

Health Benefits of Soy

In India life style diseases such as heart problems, diabetes and obesity are rising at an alarming rate. In addition to being a rich source of nutrients, soybean has a number of phytochemicals (isoflavones), which offer health benefits along with soy protein though they do not have any nutritional benefits. Soybeans contain two primary isoflavones called genistein and diadzein, and a minor one called glycitein. Isoflavones have a very limited distribution in nature. Soybeans and soy foods contain approximately 1-3 mg of isoflavones per gram. Soy protein and isoflavones together contribute to a number of health benefits such as, cancer prevention, cholesterol reduction, combating osteoporosis and menopause regulation.

For the past 30 years, investigators have shown that consumption of soy protein selectively decreases total and LDL (bad) cholesterol and maintains HDL (good) cholesterol in individuals with elevated blood cholesterol levels. In support of the various research findings, United States Food and Drug Administration issued a health claim for soy protein in October of 1999. The health claim states "consumption of 25gms of soy protein per day with a diet low in saturated fat may lower the risk of heart diseases". One year later American Heart Association endorsed the same health claim.

Soybeans are hypoglycemic and therefore very beneficial when added to a diabetic diet. The glycemic index (GI) is a numerical system of measuring how much of a rise in circulating blood sugar a carbohydrate triggers – the higher the number, the greater the blood sugar response. A diabetic person is more at a risk of getting heart diseases and because of its heart health properties soybeans is doubly advantageous for diabetic people. Soybeans glycemic index (GI) is 18-25 and it does come under the list of low GI foods, therefore good for diabetics. Its GI is lower than most legumes.

Another important aspect of soy protein is combating osteoporosis and relieving menopause symptoms. One factor in bone health is limiting the amount of calcium lost from the body. Although protein especially animal protein contributes to calcium loss, soy protein exhibits less calcium leaching effects. The isoflavones found in soybeans may also directly stop bone deterioration. Recent research has shown that soy foods can relieve most menopausal symptoms, thus reduce risks of cardiovascular disease and osteoporosis. Soybean is considered as a natural alternate for hormone replacement therapy for treating women who are in menopause stage.

Soybeans contain both soluble and insoluble fiber. Soluble fiber may help lower serum cholesterol and control blood sugar. Insoluble fiber increases stool bulk, may prevent colon cancer and can help relieve symptoms of some digestive disorders. Human epidemiological studies suggest that as little as one serving of soy foods each day may be protective against many types of cancer.

Micro Enterprise Opportunities with Soy

Soybean is an excellent crop to process at small and micro- enterprise level. The number of products that can be manufactured with a very low investment are more.

Therefore in a country like India where there is potential for innumerable small entrepreneurs this is one of the ventures which is most suitable. India has seen a major failure of soy foods in early 80's which gave a negative image to soy foods in India. Beginning early 2000, work was started working with small entrepreneurs to find out the reasons for this failure and the low popularity of soybeans as food in India. At the same time, work was started on improving the quality of existing soy food products, improving market positioning, developing new products, and creating awareness among consumers in order to realize a significant off take of soybeans in the form of food products such as soymilk, tofu, soy flour, and TVP. Work was done diligently to encourage micro, small and medium scale entrepreneurs to start entering into soy food business. Through various promotional efforts such as technical assistance and marketing promotions, soy food products started becoming popular in these markets. As a result, a number of soy food industries were started. Due to these efforts, the number of soy food manufacturers currently stands at 450 compared to about 25 in 1999-2000. These include small-, medium- and large-scale operations. Based on this success, many large food companies have also started producing and marketing various soy food products, thus raising the presence of soy food products in the marketplace to a remarkable degree. Before, the availability of soybeans was confined mostly to production areas. Currently, several of the soy food products became available in almost all the cities in India.

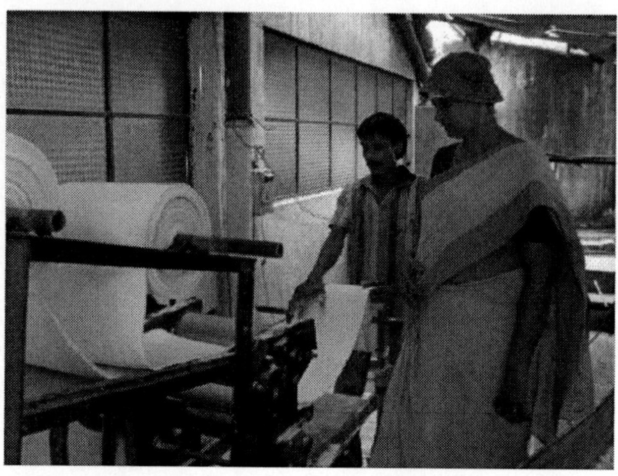

Incubation Centre

Incubation centre has been set up to see how this idea works out. It is a small center where a few basic soy product making machines are installed. These are of small capacity. Soy nut making, soy papad, soy milk and tofu etc making machines were installed. Those who do not want to take the risk of setting up of manufacturing units can use this center to make the products they want and market them. Once they understand the pulse of the market and are confident they can set up their own manufacturing unit. Instead of starting their own unit they come and use the unit in the incubation center, make the products they are interested in and then pack it under their brand and market the product. If they wish to continue they can do so and if they feel they can start their own unit they do that. This is a facility we visualised for entrepreneurs who do not have a lot of financial backing.

Women's Entrepreneurship

Women's entrepreneurship is something encouraged by all in India especially small units run be women. There are several schemes under which women are offered loans and they can set up small manufacturing units. Soy products can be one of the products women can focus. We regularly do entrepreneur training programs for women in which we focus on soy products. We talk to them about soy nutritional and health benefits, which products can be made, their market potential etc. For example, a lady who started making noodles, vermicelli and pasta incorporating soy and she has products which she sells as gluten free products.

She is one of the few lady entrepreneurs who have successfully started their units and are running them.

Soy Protein Products

The defatted soy flakes remaining after oil extraction are the basis of a variety of soy products including soy flour, soy concentrates and soy isolates. Soy concentrates contain about 65% protein and retain most of the bean's dietary fiber. Concentrates when added to food products enhance texture and help foods retain moisture. Soy protein concentrates are used in products such as Surimi, soup bases and gravies. Soy protein isolate has about 90 per cent protein, very little moisture with no fiber, carbohydrates or fat. Soy isolates are the chief components of many dairy-like products, including soy based cheese, soymilk, infant formula, non-dairy frozen desserts and coffee whiteners. Soy protein products are used to add texture to meat products and are valued for their emulsifying properties. Protein supplements (nutraceuticals) are a group of products in which soy protein-isolate is a major component and these are used in weight reduction programs, body building regimes and during convalescence.

Fortification of Wheat Flour

Wheat-soy flour is the wheat flour fortified with defatted soy flour maximum up to 10 per cent. Protein content in What-soy flour is about 16-17% and there will be about 25-30 percent increase in protein content compared to regular whole-wheat flour. Soy flour is rich in protein and has an excellent profile of amino acids, particularly lysine, which is deficient in wheat flour. Blending of 10 % soy flour with wheat flour will not only provide higher protein content but also improve the amino acid balance. Research shows that, blending of soy flour with wheat flour will increase the recommended amino acid availability from 40 to 80%.

In addition to nutritional improvements, soy fortified wheat flour will improve the functional characteristics of the end products in terms of better moisture retention and lower oil absorption.

Soy Fortified Gram Flour

Defatted Soy Flour (DSF), which has more than double the protein of chickpea flour, is cheaper than chickpea flour. Chickpea flour is used widely in India to make innumerable snack and other products. At the same time, DSF blends very well with *besan* with no beany flavor. About 20% of *besan* can be replaced with defatted soy flour without affecting the organoleptic attributes. Substitution of *besan* with defatted soy flour (20%), results in increase in the protein content from 20 to 26% in the blend translating to an increase of 30% in protein content and more than 30% improvement in protein quality. Extensive research has been done on soy fortified *besan*-based traditional products. This research shows that fortification of *besan* with soy flour has a number of advantages, specifically, nutritional improvement, cost effectiveness, a key functional benefit (less oil absorption) and overall health benefits. Key functional properties – fat binding capacity and water absorption capacity – are quite different for soy protein compared with gram flour. It has been observed that addition of 20% soy flour to gram flour almost reduced 20% reduced the oil absorption during frying.

Soymilk and Tofu

Soymilk is an aqueous extract from soybeans. When extracted from soybean by using modern technologies soymilk can be made to taste great while containing all the nutrition of soybean. Soymilk can be handled and used much in the same way as dairy milk. Besides being rich in protein, vitamins and minerals, soymilk is lactose free, cholesterol free and low in saturated fat. It presents a highly nutritious alternative to dairy milk and very beneficial for lactose intolerant people. Besides making tofu, soymilk can be converted into hot and cold beverages, fruit shakes, yogurt and ice cream. Soymilk is available in bottles, poly packs and bulk

also. Bulk and poly packed soymilk needs to be refrigerated. Bottled milk is usually sterilized and does not require refrigeration but it should be served chilled for best taste. Tofu (soy *Paneer*) is the most popular among all the soymilk products. It is made upon coagulating the hot soymilk and removing the whey. Tofu is a versatile food and can be converted into a variety of value added products. Tofu is bland easily takes the flavor of the product with which it is cooked.

Soy Nuts

Soy nuts are whole soybeans that have been soaked in water and roasted until brown. Soy nuts, like whole soybeans, are excellent source of protein, fat and isoflavones. Soy nuts can be eaten as snack. They can be manufactured either by dry roasting or deep-frying in oil. They can be eaten as an alternate to peanuts, which are expensive, and pose the problem of aflatoxins. Most conventional nuts are incredibly high in fat but soy nuts have less fat and more protein compared to conventional nuts. Roasted soy nuts have at least 50% more protein and 50% less fat than peanuts. Soy nuts are similar in texture and flavor to peanuts and are far less expensive than peanuts.

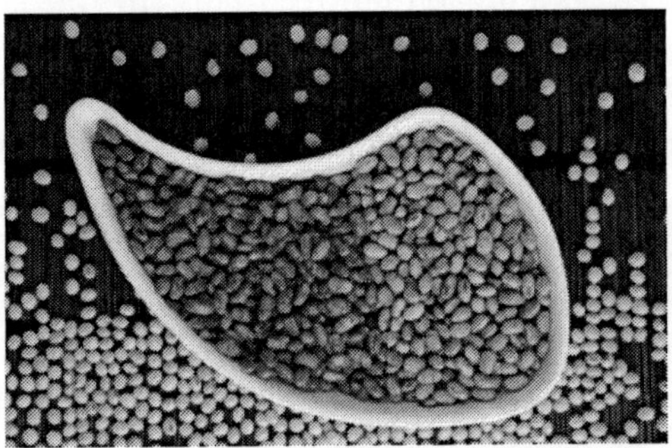

Nuggets and Granules

Nuggets and granules, technically termed Texturised Vegetable Protein (TVP), are economical and convenient for consumption for all consumers. TVP is made from a very good quality defatted soy flour using a highly versatile extrusion

technology. These products have more than 50 percent protein, which makes it one of the richest sources of protein. Its nutritional composition is similar to that of defatted soy flour. It is ideal as a protein source in a vegetarian diet. These products have hydration properties that are ideal for a wide range of food applications. TVP can be used to replace vegetables and meat in a variety of recipes. The product is bland and blends itself with the other ingredients to which it is added. In addition to the health benefits, nuggets and granules are low in fat and a good source of iron and B-vitamins. As they are pre-cooked, they are easily digestible and require very less time to cook. The extrusion cooking process destroys the anti-nutritional factors associated with soy completely. Since the extrusion process is a high temperature short time process the nutrient loss is minimal. Nuggets and granules are very low in moisture so can be kept for long time.

Bakery Products

In different bakery products wheat flour can be fortified with soy flour both for functional and nutritional reasons. In general soy flour can be added up to 3% without changing formulation except water. Currently a number of bakers are using enzyme active soy flour up to 1% as a bread improver. This enzyme active soy flour can be either full fat or defatted. The enzyme lipoxigenase present in soy not only improves the bread color but also improves machinability (dough handling properties) of the dough. In case of bread and biscuits, high Protein Dispersability Index (PDI) soy

flour (low heat treated) can be used as an alternate to non-fat dry milk, which can reduce the cost of the raw material. Addition of soy flour improves the nutritional quality as well as increases water absorption, helps in emulsification of fats and other ingredients. The other major application for soy flour in bakery products is use of lecithinated soy flour as an alternate to whole eggs in cakes and doughnuts. High PDI soy flour with 6-15 % of lecithin can be used in cakes and doughnuts to replace whole egg in these products. A recent study conducted at American Institute of Baking in the U.S. suggested that by using lecithinated soy flour as an alternate to egg, the raw material cost could be almost reduced by 25%.

Noodles

Noodles, pasta and vermicelli are becoming extremely popular in India. These products are prepared by means of an extrusion machine that is basically made of stainless steel. The machine provides the manufacturers a choice to produce these products with several alternative raw materials (like white flour, semolina, rice flour and so on) and of different types (like Spaghetti, Fettuccini, Macaroni, Fusilli, Penne, etc.) of pasta and noodles. The basic raw material for noodles is, refined wheat flour, which has inferior nutritional quality. Addition of soy flour will not only improved the nutritional quality it also gives some functional benefits to the final product. Addition of 10-12% defatted soy

flour to wheat flour will give the product increased moisture retention in case of wet noodles. Where as, in case of fried noodles it will save oil during frying. Soy fortified noodles not only increases the noodles strength but also increase in the product yield without any effect on color and taste. The improved noodle strength reduces the breakage losses during drying of noodles and requires no oil for handling of the noodles especially to separate the wet noodles. Just addition of 10% soy flour almost reduces 60 % usage of eggs in case of egg noodles.

Conclusion

Soybeans supply protein, minor components, and fiber together with cholesterol-reducing and anti-carcinogenic properties. Awareness of soybean's properties and of the use of soy protein in foods has increased dramatically during the last few years. As consumer demand grows for low-fat healthy foods, the use of soybeans as a key ingredient can be expected to expand. In most of the products, except wheat products, addition of soy flour generally reduces the cost and improves the functionality. Where as, in wheat products even though there is a slight increase in cost, if the functionality of soy protein in various products is considered, addition of soy flour is economical with nutritional improvement as bonus.

OO

Changing Scenario of Food Processing Industry

Dheer Singh and D K Bhatt

Food industry is one of the most emerging industries in the present scenario. India is the world's second largest producer of food next to China, and has the potential of being the biggest with the food and agriculture sector. The food processing Industry is one of the largest in India – it is ranked fifth in terms of production, consumption, export and expected growth. The food industry is on a high as Indians continue to have a feast. The demand for food and specially the processed food is increasing day by day. What will the future hold as the world population grows from 6.5 billion today, to an estimated 10 to 12 billion by 2030. Along with the increased demand in food quality, futurists predict a demand for an improvement in the quality of food. How will we meet future human food needs, both in quantity and quality? The only way is more research for higher production and better processing/preservation. This is possible only by taking the world closer through collaboration in research and presentation of work in the world forum. Different countries have different production and processing share in the world scenario.

Food grain production has touched 230 Mt in 2008-09 from a mere 51 Mt in 1951-52. The country has attained self sufficiency in food grain production and also have sufficient buffer stock. The country has also made impressive progress in production of food grains, oilseeds, horticultural crops, milk, poultry, etc

India has major scope and opportunities in the food processing industry. The Indian food processing market is one of the largest in terms of production, consumption, and export and import prospects. The total food production in India is likely to double in the next ten years and there is an opportunity for large investments in food and food processing/preservation technologies, skills and equipment, especially in areas of Canning, Dairy and Food Processing, Specialty Processing, Packaging, Frozen Food/Refrigeration and Thermo Processing. Fruits & Vegetables, Fisheries, Milk & Milk Products, Meat & Poultry, Packaged/ Convenience Foods, Alcoholic Beverages & Soft Drinks and Grains are important sub-sectors of the food processing industry. India is one of the world's major food producers but accounts for less than 1.5 per cent of international food trade.

This indicates vast scope for both investors and exporters. Indian Food and Beverages Forecast (2007-2011)" report gives an in-depth analysis of the present and future prospects of the Indian food and beverages industry. It looks into the industry in detail with foci on organized food retailing, consumer food purchasing behaviour, food processing industry and packed/convenience food industry. The report states that Supermarket sales will expand at a much higher rate than other retail formats.

This is because greater number of higher income Indians will prefer to shop at supermarkets because of convenience, higher standards of hygiene, and attractive ambience. It is also expected that fruit consumption will increase at a CAGR of 4.33% for the period spanning from 2007-2011, highest among all the food products taken in this report. The report further goes on to say that the processed-food market is the main focus for foreign companies as this segment is underdeveloped and presents enormous potential for growth. The growth of modern, organized retailing—in contrast to the kiosks and small shops from which Indians have been purchasing food traditionally—will also increase the demand for value-added foods.

Increasing incomes are always accompanied by a change in the food basket. The proportionate expenditure on cereals, pulses, edible oil, sugar, salt and spices declines as households climb the expenditure classes in urban India while the opposite happens in the case of milk and milk products, meat, egg and fish, fruits and vegetaables.

The proportionate expenditure on staples (cereals, grams, pulses) declined from 45 percent to 44 percent in rural India while the figure settled at 32 per cent of the total expenditure on food in urban areas.

A large part of this shift in consumption is driven by the processed food market, which accounts for 32 per cent of total food market. It accounts for US$ 29.4 billion, in a total estimated food market of US$ 91.66 billion.

The Confederation of Indian Industries (CII) has estimated that the foods processing sector has the potential of attracting U$ 33 billion of investment in 10 years and generate employment of 9 million-days.

The government has formulated and implemented several plan schemes to provide financial assistance for setting up and modernizing food processing units, creation of infrastructure, support for research and development and human resource development in addition to other promotional measures to encourage the growth of the processed food sector.

Food processing is a large sector that covers activities such as agriculture, horticulture, plantation, animal husbandry and fisheries. It also includes other industries that use agriculture inputs for manufacturing of edible products. The

Ministry of Food Processing Industries, Government of India indicates the following segments within the food processing industry:

Meat and Poultry processing

Grain Processing

Dairy Products

Fruits and Vegetables Processing

Fisheries

Consumer foods including packaged foods, beverages and packaged drinking water.

Though the industry is large in size, it is still a nascent in terms of development of the country's total agricultural and food produce, only 2 percent is processed. The industry size has been estimated at US$ 70 billion by the Ministry of Food Processing Industries, Government of India.

Value addition of food products is expected to increase from the current 8 percent to 35 percent by the end of 2025. Fruits and vegetables processing, which is currently around 2 percent of total production will increase to 25 percent by 2025.

The highest share of processed food is in the dairy sector, where 37 percent of the total produce is processed, of the only 15 per cent is processed by the organized sector. The food processing industry in the country is on track to ensure profitability in the coming decades. The sector is expected to attract phenomenal investment of about Rs. 1,400 billion in the next decade.

In the dairy sector, most of the processing is done by the unorganized sector. Though the share of organized sector is less than 15 per cent, it is expected to rise rapidly, especially in the urban sector. Nestle and Britannia, Multi National Corporations, have forayed into emerging segments such as Ultra Heated Treatment (UHT) and flavour milk. UHT milk is becoming popular and the market is estimated at US$ 33.4 million.

India produces more than 200 million tonnes of different food grains every year. All major grains – rice, wheat, maize, barley and millets like jowar (great millet), bajra and ragi are produced in the country. About 15 per cent of the annual production of wheat is converted into wheat products. There are 10,000 pulse mills in the country with a milling capacity of 14 million tonnes, milling about 75 per cent of annual pulse production. India is the second largest rrice producer in the world after China with 20 percent global share.

India has a live stock population of 470 million, which includes 205 million cattle and 90 million buffaloes. Total meat production in the country is currently estimated at 5 million tonnes annually. Only about 1-2 per cent of the total meat is converted into value added products. The rest is purchased as raw and consumed at home. Poultry processing is also at a nascent stage. The country produces about

450 million broilers and 33 billion eggs annually. Growth rate of egg and broiler production is 16 percent and 20 per cent respectively.

The growing number of fast food outlets in the country has had a significant impact on the meat processing industry in India. Most of the production of meat and meat products continues to be in the unorganized sector. Some branded products like Venky's and Godrej's Real Chicken are, however, becoming popular in the domestic market.

India is the third largest fish producer in the world and is second in Inland fish production. The fisheries sector contributes US$ 4.4 billion to the national income, which is about 1.4 percent of the total GDP. With its over 8,000 km of coast line, 3 million hectares of reservoirs, 1.4 million hectares of brackish water, 50,600 sq km of continental shelf area and 2.2 million sq km of exclusive economic zone, India is endowed with rich fishery resources and has vast potential for fishes from both inland and marine resources. Processing of fish into canned and frozen forms is carried out almost entirely for the export market. It is widely felt that India's substantial fishery resources are under utilized and there is tremendous potential to increase the output of this sector.

Consumer food is a big segment which includes packaged foods, aerated soft drinks, packaged drinking water and alcoholic beverages. The packaged or convenience food are making a big share in the food market. This segment mainly comprises of bakery products, ready to eat snacks, chips, namkeens (salted snacks and savouries) and other processed foods. The market size of confectionaries is estimated at US$ 484.3 million and growing at the rate of 5.7 per cent per annum.

In the aerated soft drink sector two of the biggest global brands in this segment are well established in India. Soft drinks constitute the third largest packaged foods segment, after packed tea and packed biscuits.

So there is a vast scope for the growth of food processing industry as India is having a small share in the world processed food market.

8

Influence of Optimal Harvest Stage and After-Ripening on Seed Quality of Processing type Hot Pepper (*Capsicum annuum L.*)

Rakesh C Mathad, S.B. Patil, S.N. Vasudevan and B.L. Lokeshappa

Abstract

India is the second largest producer of vegetables after China and a major destination for food processing industries which is fifth biggest in the world. Value addition to perishable vegetables through food processing generates more income and employment. Among many vegetables hot pepper is a major vegetables being used for processing. Many products like sauces, chutneys, powders, oleoresin, pickles etc are made out of hot pepper. Unlike in tomato where there are varieties available specifically for processing in hot pepper the development of cultivars and standardization of seed production technology is to be done. There are very few processing type varieties available in hot pepper. In an experiment conducted in UAS, Raichur in 2011 to know the influence of optimum harvest stage and after-ripening 6 commercially available processing type hot pepper varieties were evaluated for seed quality. Among the 6 varieties 3 are Asian type and 3 are Exotic type. The Asian varieties T-236, N-765 and R-118 performed well with respect to seed quality parameters like average good seed / fruit, 1000 seed weight, germination % and Usable transplant % when harvested at 2-weeks after pollination and 5-days of after-ripening. Whereas the exotic type varieties S-586, X-770 and H-768 showed higher seed quality parameters when harvested at 4-weeks after pollination and 2-days of after-ripening. Both the groups of varieties were harvested at 98% colour development stage. The seeds were analysed under image microscopy and various stages of seed development like seed filling, embryo development, ageing, mechanical damages were documented at different harvesting stages.

India is the second largest producer of vegetables, with estimated 135 MT from 6.5 MHa of land. This huge production can be achieved with the use of improved seed and technology. India is having unique distinction of being a major seed production hub of Asia. With the exception of some brassica seeds most of vegetable seeds produced in India. The higher vegetable production help the food processing industry in India which is among the fifth largest in the world. In vegetables like tomato there are specific processing varieties available for food processing. In other vegetables the varieties specifically for food processing are yet to be evaluated. Though there are some asian and exotic (bred for European conditions) are available the seed production technology is yet to be standardized. The optimal harvest stage and after ripening are most important criteria for harvesting quality seed. To find the stage at which the seed to be harvested and effect of after ripening in commercially available processing type hot pepper is the main objective of this paper.

Methods

A experiment was undertaken in UAS, Raichur in 2011 to know the optimal harvest stage in hybrid hot pepper using 6 commercially available processing type varieties namely T-236, N-765, R-118 (asian type), S-586, X-770 and H-768 (exotic type). The seeds of these varieties were obtained from the leading companies producing vegetable seeds operating in India.

The methods used are tabulated as follows:

Growing conditions	Noted are sowing, planting and pollination (start + finish) + harvest date. The data on Temperature and Relative Humidity were collected using a data logger and down loaded in to a computer. The nutrition level of the soil / substrate is determined 4 weeks after start pollination on PH, E.C., N, P, K, Ca, P, Mg and micro elements.
Sampling	First time in the second harvest week (cluster 2 –3) after pollination. The second time a few weeks later, at 4th week before finish crop.
Harvest	10 Orange /red fruits of colour stage 7 or 98% colour development are taken at random from block 1 & 2 each. Only healthy fruits were harvested and there may be no selection on fruit size or shape.
Weighing fruits	The fruit weight of the 10 fruits per treatment are determined. (0 or 1 decimal) at day of harvest. Each sample get a code: name – yng/old – colour – days-after ripening – replication.
After-ripening	The after ripening is done at around 20-25ºC (in the shade !!!) for exact the requested no. of days i.e. 2 or 5 days.
Extraction	Seed extraction is done in the end of the afternoon. The seeds may not be damaged. The fruits are cut open and the seed with the pulp are taken out. Then seeds washed with clean running water before dried in a dryer. No seed treatment of what so ever is done.

Remove free water	by centrifuge or with paper
Drying	put the seeds on a dryer, or in a thin layer on paper. Temperature of **20 to 25 °C.** The seeds must be **dry within 12 hr**. Use a little ventilation and stir the seeds **gentle after 2 and 4 hr**. to avoid double seeds and for some polishing. This stirring is important it avoid a lot of work for later separation of double seeds + less damage.
Storage	Store the seeds cool (< 15ºC) and dry (< 40% RH / 7% M.C.) till sending to lab for micro analysis.
Tests	TZ-test
Cleaning	Rub the seeds carefully with your hand for some polishing and to detach the last double seeds. With a blower the empty and light seeds are removed.
Sizing	With round sieve 2.25. Count and weigh the seeds of both sizes of size > 2.25
Germination	Both soil germination with 4 X 40 seeds (un-treated) and between the paper germination test with ISTA approved methods
Image Microscopy	The seeds of various varieties analysed under image microscopy to know the various developmental stages.

Figure 8.1: *Good Seed-T-236 (Asian) at 98% 2W:DAF+5 Days AR*

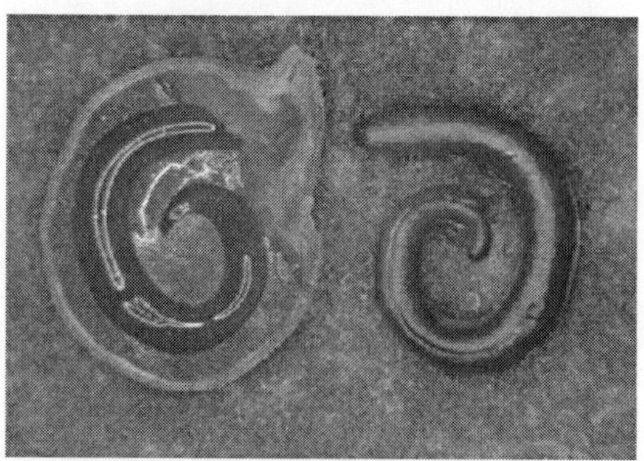

Figure 8.2: *Good Seed-X-770(Exotic) at 98%-4W:DAF+2 D- AR*

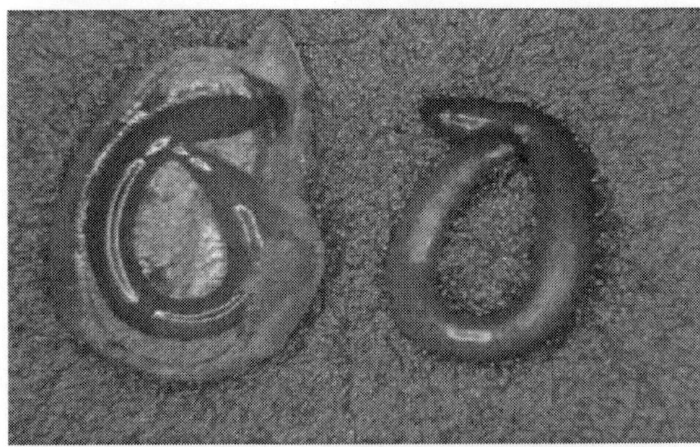

Figure 8.3: *Good Seed-R-118 (Asian) at 98%-2W:DAF+5 Days AR*

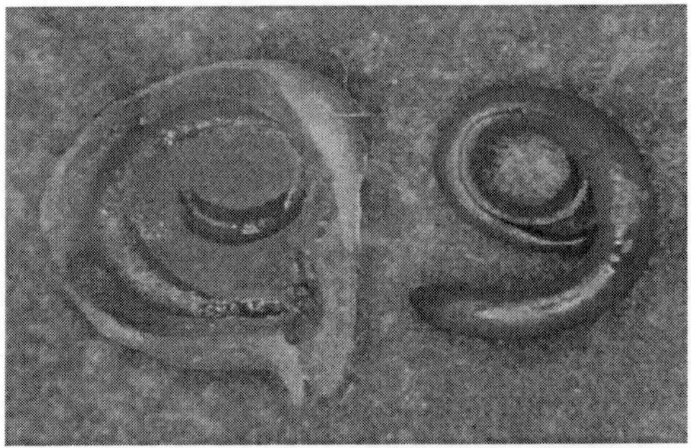

Figure 8.4: *Good Seed-H-768 (Exotic) at 98%-4W:DAF+2 D-AR*

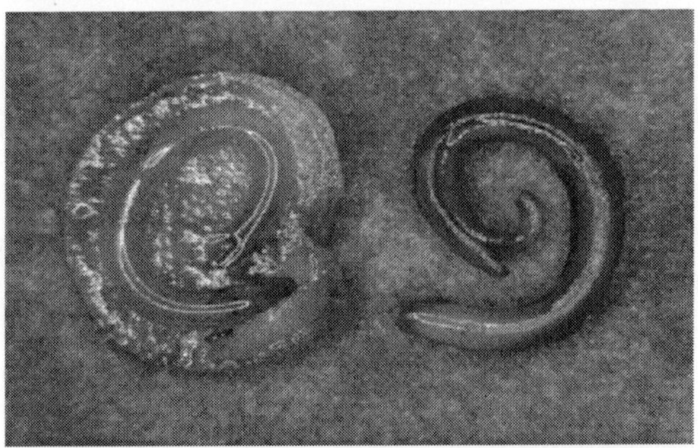

Figure 8.5: *Overripe in Exotic Type Varieties*

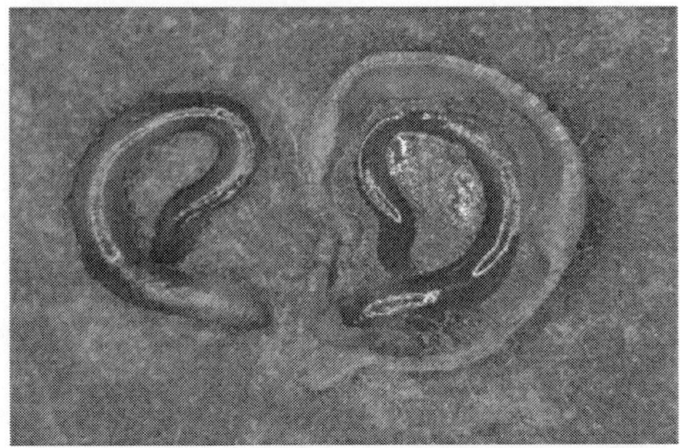

Figure 8.6: *Pregerminated-in Asian Type Varieties*

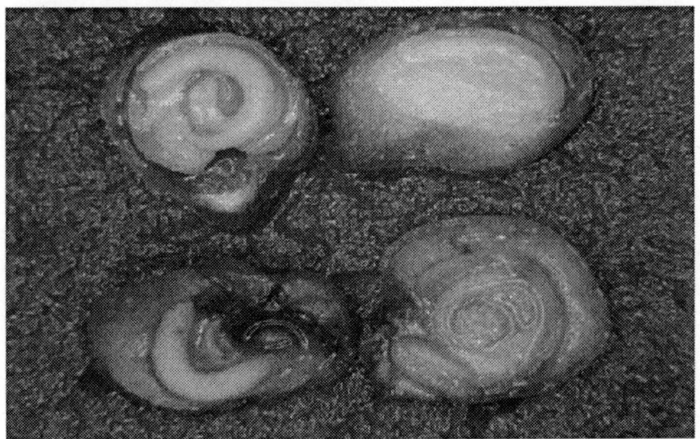

Figure 8.7: *Poorly filled in Asian Type Varieties*

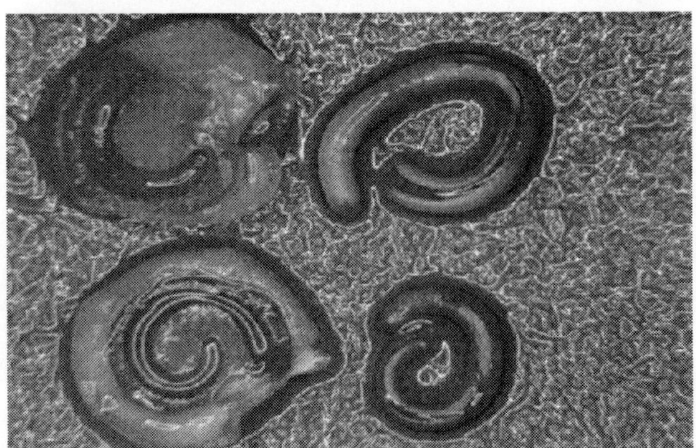

Figure 8.8: *Poorly filled in Exotic Type Varieties*

TABLE 1A

Effect of Optimal Harvest Stage and After Ripening on Germination & UT %

Treatments / Varieties	*GERMINATION %*						*Av.*	*USABLE TRANSPLANT %*						*Av.*
	Asian Proc. Type			*Exotic Proc. Type*				*Asian Proc. Type*			*Exotic Proc. Type*			
	T-236	*N-765*	*R-118*	*S-586*	*X-770*	*H-768*		*T-236*	*N-765*	*R-118*	*S-586*	*X-770*	*H-768*	
98 % Col+2 WAP+2 AR	89	90	90	92	95	95	91.83	88	79	89	88	89	88	86.83
98 % Col+2 WAP+5 AR	94	94	96	89	93	92	93	90	91	91	90	79	88	88.17
98 % Col+4 WAP+2 AR	92	91	89	97	97	95	93.5	87	88	76	92	92	91	87.67
98 % Col+4 WAP+5 AR	90	89	88	86	93	94	90	83	86	76	85	89	83	83.67
Mean	91	91	91	91	95	94	92.0	87	86	83	89	87.2	87.5	86.5
CD	(P-0.05)							(P-0.05)						
Harvest Stage	1.65							1.99						
Varieties	1.35							1.63						
Interaction	NS							NS						

TABLE 1B

Effect of Optimal Harvest Stage and After Ripening on 1000 Seed Weight & GSF %

Treatments / Varieties	*1000 Seed Weight*						*Av.*	*Good Seed per Fruit %*						*Av.*
	Asian Proc. Type			*Exotic Proc. Type*				*Asian Proc. Type*			*Exotic Proc. Type*			
	T-236	*N-765*	*R-118*	*S-586*	*X-770*	*H-768*		*T-236*	*N-765*	*R-118*	*S-586*	*X-770*	*H-768*	
98 % Col+2 WAP+2 AR	5.94	5.41	5.78	7.09	6.09	5.23	5.92	95	87	82	88	86	74	85.33
98 % Col+2 WAP+5 AR	6.09	5.56	6.06	7.17	6.08	5.33	6.05	95	92	93	89	89	80	89.67
98 % Col+4 WAP+2 AR	5.55	5.23	5	7.44	6.34	5.64	6.15	93	88	83	94	88	89	89.17
98 % Col+4 WAP+5 AR	5.37	5.2	5.11	6.9	6.09	5.23	5.65	89	88	78	82	79	86	83.67
Mean	5.73	5.35	5.48	7.15	6.15	5.35	5.94	93	88	84	88	85	82	86.96
CD (P-0.05)	(P-0.05)							(P-0.05)						
Harvest Stage	0.24							2.03						
Varieties	0.19							1.66						
Interaction	NS							NS						

Results and Conclusion

The results mainly with reference to seed quality parameters like average good seed/fruit, 1000 seed weight, germination % and Usable transplant % were analysed. The data on usable transplants was taken since germination is not exact measure of no. seedlings we are going to get when sown in the seed bed or nursery. Since hot pepper is high value seed measuring usable transplants along with germination is accurate method of standardization of harvesting stage. The results show that the asian type varieties T-236, N-765 & R-118 harvested at 98% colour development-2 Weeks After Pollination & 5 Days of After Ripening have recorded highest germination (94, 94 & 96), Usable Transplant % (90, 91 & 91), 1000 Seed Weight (6.09, 5.56 & 6.06) and Good Seed per Fruit (95, 92 & 93). (Table 8.1A & 8.B). The image microscopy also shown good well developed nicely filled good seed as seed in figure 8.1 & 8.3. The early harvesting of seeds shown poor filling of seeds (Figure 8.7). The delayed harvesting shown pre-germination of seeds inside the fruit (Figure 8.6).

The exotic type varieties S-586, X-770 & H-768 harvested at 98% colour development-4 Weeks After Pollination & 2 Days of After Ripening have recorded highest germination (97, 97 & 95), Usable Transplant % (92, 92 &91), 1000 Seed Weight (7.44, 6.34 & 5.64) and Good Seed per Fruit (94, 88 & 89). (Table 8.1A & 8.1B). the image microscopy also shown well developed seed structures as in figure 8.2 & 8.4. The seeds harvested beyond this stage shown poorly filled and over ripen seed which gave low germination % (Figure 8.5 & 8.8).

References

Alan, Eser. B, 2008. *The Effect of Fruit Maturity and Post Harvest Ripening on Seed Quality in Hot & Conic Pepper Cultivars.* Seed Science and Technology., 36 (2): 467-474.

Victor, M. Sanchez, F.J. Sundstrom, G.N. McClure, N. S. Lang., 2003. *Fruit Maturity, Storage and Post-harvest Maturation Treatment Effect on Bell Pepper Seed Quality.* Scientia Horticulturae, 54 (3): 191-201

Demir, I. and R.H. Ellis, 1992. *Development of Pepper* (Capsicum annum L.) seed quality. Ann. Applied Biol., 121: 385-399.

Marcelis, L.F.M. and L.R. Baan-Hofman-Eijer, 1995. *Growth Analysis of Sweet Pepper Fruits* (Capsicum annuum L.). Acta Hortic., 412: 470-478.

Raymond, A.T.G., 1999. Vegetable Seed Production. CABI Pub., Wallingford, Oxon, UK., pp: 231-235.

Bangerth, F., 1989. *Dominance among Fruits/Sinks and the Search for a Correlative Signal.* Physiol. Plant, 76: 608-614.

○○

A Pyroaurite Type Sorbent for Removal Arsenate from Water by Column Process

Shalini Shrivastav, Ranjeet Varma and V.P.S. Bhadauria

Abstract

In an attempt to evaluate the suitability of synthesized Pyroaurite Like Compound (PLC) and modified PLC, i.e., Pyroaurite Type Sorbent (PTS) for the removal of pentavalent arsenic (arsenate) from water, a comparative study of these two adsorbents was carried out by column process. Arsenic was analyzed using Perkin-Elmer AAnalyst 100 AAS using MHS system. In column tests, PLC and PTS could remove 60.5 and 64.8% As(V) respectively at initial metal concentration 0.1 mg/l, pH 6.0, temperature 25°C, bed height (BH) 60 cm and flow rate 1.0 ml/min. The order of arsenate removal with these chemical adsorbents was found to be PTS > PLC. The effects of various parameters affecting the adsorption efficiency such as initial arsenate concentration, bed height, temperature and pH were studied.

Introduction

The occurrence of arsenic in natural ground waters is usually due to mineral arsenopyrite, which is associated with sedimentary rocks and weathered volcanic rocks. Fossil fuel, mineral deposits, mining wastes and geothermal areas are other sources through which arsenic can occur in groundwater (Ferguson and Gavis, 1972). Also, arsenic and its compounds are found in wastes from various industries like glassware, ceramic dyes petroleum and refining, metallurgical, insecticides, pesticides, fertilizers and inorganic chemicals (Blakely, 1984). Thus, a huge natural occurrence along with manmade sources are responsible for increasing arsenic concentration in groundwater and is causing serious problems as it is toxic even at a very low concentration. Symptoms of arsenic poisoning are of various types e.g. dermatological lesions, liver disorder, muscular weakness and paralysis of lower limbs, etc. (Saha et al., 1999). It may cause neurological damage if arsenic is greater than 0.1 mg/l; if concentration is greater than 9-10 mg/l, it may cause impairment of bone marrow function and neurological abnormalities and skin cancer (Nic and Korte, 1991). Arsenical dermatosis was detected first time in India in 1983, in 14 villages in 5 districts of West Bengal (Chakarborti and Saha, 1987). The usual manifestations were diffuse melanosis and keratosis. In West Bengal few arsenic patients died, while suffering from internal cancers i.e. lung, bladder, liver cancer etc., arsenic melanosis was the most common symptom among the arsenic victims.

In the United States, the Environmental Protection Agency has proposed lowering the maximum contaminant level for arsenic in drinking water from 50 to 10 μg/l, but the feasibility of the proposed standard is currently being evaluated. The European maximum admissible concentration and the World Health Organizing guideline for arsenic in drinking water are both set at 10 μg/l. On the other hand, developing countries are struggling to find and implement measures to reach of 50 μg/l in arsenic affected areas.

Various methods of treating arsenic available today are coagulation, precipitation, ion-exchange and adsorption (Nagarnaik et al., 2004; Raichur and Panvekar, 2002). Removal of arsenic using various adsorbents (Neeri news, 2004) like chitsin (Elson et al., 1980), sand (Vaishya and Agarwal, 1993) basic yittrium carbonate (Wassay et al., 1996) and coconut husk carbon (Manju et al., 1998) is studied. So, to evaluate the suitability of synthesized Pyroaurite Like Compound (PLC) and modified PLC, i.e., Pyroaurite Type Sorbent (PTS) for the removal of pentavalent arsenic (arsenate) from water, a comparative study of these two adsorbents was carried out by column process.

Materials and Methods

As(V) Solutions

All reagents were of analytical grade. In order to avoid interference by other elements in wastewater, the experiments were conducted with aqueous solutions of pentavalent arsenic in redistilled water. Synthetic samples of various initial As(IV) concentrations were prepared by dissolving AR grade sodium arsenate (Na_2HAsO_4) in water. Four standard solutions of 0.1, 1, 5 and 10 mg/l concentrations of As(V) were prepared by diluting the stock solution of 1 g/l that was made by dissolving 3.636 g $Na_2HAs\,O_4$ in double distilled water.

Preparation of Pyroaurite like Compound

A pyroaurite like compound (PLC) was synthesized from $MgCl_2$ and $FeCl_3$ following a reported procedure (Seida et al., 2001).

Immobilization of PLC onto Cellulose Acetate: Clay type compounds such as PLC cannot be used in columns because they coagulate among the particles and swell due to their water absorbing property. So PLC was immobilized on a hydrophobic support of cellulose acetate.

Sorbent Characterization: PLC and modified PLC, i.e., PTS were analysed for their physicochemical characteristics and chemical analysis such as density (g/cc), bulk density (g/ml), moisture (%), ash (%), volatile matter (%), combustible matter (%), loss on ignition (%), SiO_2(%), Al_2O_3 (%), CaO (%), Fe_2O_3 (%), MgO (%) and surface area (m^2/g).

The bulk or apparent density of soil or sorbent is its mass per unit volume. It was obtained by dividing mass by volume of a substance. The sorbent filled upto the brim in weighing bottle was weighed and its volume was found by filling the bottle fully with water and measuring the volume of the water. The bulk density of most soils range from 1.02 to 1.8 g/ml. The true or particle is the average density of the particles of dried sorbent. It is obtained by the RD method in which the mass of sorbent is divided by the mass of water displaced. For soils it is generally 2.65 g/cc, i.e., the specific gravity is 2.65. Porosity of sorbent is the fraction of sorbent not occupied by sorbent particles. The bulk density is less than true density because porosity is included in the bulk. Moisture content of the sorbents was determined by weighing about 1 g of sorbent in silica dish before and after drying for 2 h in an oven at 110°C. After determining the percentage of moisture, the content in the dish was ignited at a bright red heat for 30 min to calculate the loss of ignition. Surface area was determined by the glycol retention method.

To find the total mineral constituents of a soil or sorbent, it is digested with acids such as HCl, HNO_3+H_2SO_4+$HClO_4$ called triple acid etc. or fused with sodium carbonate. The most popular method, extraction with HCl was followed here. Al_2O_3 (%) and Fe_2O_3 (%) were determined using 2% NH_4NO_3 solution and igniting the precipitate gently. CaO percentage was found by precipitating calcium oxalate from the HCl extract and then titating it with N/20 $KMnO_4$ solution. MgO percentage was estimated by precipitating Mg as $MgNH_4PO_4$ from the HCl extract and then igniting it to get MgP_2O_7 (pyrophosphate). SiO_2 percentage was determined by fusing the sorbent with sodium carbonate, decomposing the fused mass with with HCl and evaporating the solution to get gelatinous silicic acid $SiO_2.xH_2O$ as insoluble silica $SiO_2.yH_2O$. The residue was extracted with dil. HCl, filtered and the residue was ignited in a Pt crucible at about 1050°C to SiO_2.

Sorbent Modification: ***(A) Sulphidation:*** Mediated metal sulphides MnS, FeS, CuS, ZnS and $CuFeS_2$ were prepared from metal salts preferably nitrates using $Na_2S.9H_2O$, Cetyl Trimethyl Ammonium Bromide (CTAB) and ethanol. The surfactant mediates the particle size of sulphides, resulting in a smaller particle size and the greater surface area for a greater sorption. The mediated metal sulphides were activated by stirring them with saturated aqueous solution of oxalic acid at room temperature. This is because carboxyl groups are, in many cases, largely responsible for sorption of cations. Activated sulphides were used to sulphidate PLC.

(B) Thiolization

I. *Monothiolation* to prepare PLC-DT: To form small functionalized spheres, dodecan-1-thiol (DT), RH, PLC and CTAB were mixed and treated as per synthesis details for given by Nooney et al. (2001) for fuctionalizing mesoporous silicas with 3-mercaptopropyl-trimethoxysilane (MTMS).

II. *Dithiolation* to prepare PLC-DTD: In the above procedure DT were replaced by dodecyldithiodecane ($C_{12}H_{25}$—S—S—$C_{12}H_{25}$).

III. *Silylation* to synthesize PLC-MTMS: A one-step silylation process were followed as Lagadic et al. (2001) did to prepare Mg-MTMS.

(C) Quaternization

A 1:1 mixture of PLC and RH were quaternized following the Lee method (1999).

(D) Physical Modifications

The crushed PLC and PTS were pelletized separately by injecting them into a commercial pellet mill. Steam was used if the moisture content was low, but no other binders or additives were used. The pelletizing process improves the structural and physical properties of sorbents making them more suitable for use in packed columns.

Column Process: In order to establish the usefulness of PTS for in-line removal of toxics from wastewater, column studies were conducted. Continuous downflow experiments were conducted using PVC column (100 cm height × 2.5 cm i.d.). Five columns were placed in series that could provide different Empty Bed Contact Times (EBCTs of 10, 20, 30 and 50 min) and bed heights (BHs of 15, 30, 60 and 75 cm). The EBCT is the average residence time and is obtained by dividing the bed volume including voids by the flow rate. The columns were inclined differently to vary flow rates naturally without pumping. All the columns were packed with previously analyzed sorbent on a small plastic wiremesh support. To study the effect of influent concentration 100 ml of each 0.1, 1, 5 and 10 mg/l As(V) solutions were passed at the rate of 1.0 ml/min through columns each having the BH 15 cm at a pH of 7.2 ± 0.2. The effect of flow rate was investigated using a column of 75 cm BH and varying the flow rate from 1.0 to 5.0 ml/min. The effect of pH was studied by changing the pH of the feed from 2 to 8 with 0.1M HCl and 0.1M NaOH. To determine exhaustive capacity in all the experiments the eluant was collected after breakthrough and analyzed for the toxic. The initial amount of the metal in 20 ml feed minus the amount found in the effluent gave the amount of the toxic retained by the sorbent. The process was continued until the amount of toxic is same in 20 ml of the feed and the effluent. The % amount of toxic removal was calculated.

TABLE 9.1

Characteristics of Adsorbents

Characteristics	*PTS*	*PLC*
Density (g/cc)	2.20	2.50
Bulk density (g/ml)	1.6	1.8
Moisture (%)	11.31	16.31
Ash (%)	9.65	12.65
Volatile matter (%)	6.74	5.74
Combustible matter (%)	64.16	54.16
Loss of ignition	5.02	4.02
Chemical AnalysisSiO_2 (%)	1.99	1.99
Al_2O_3 (%)	0.03	0.03
CaO (%)	0.11	0.11
Fe_2O_3 (%)	5.31	5.01
MgO (%)	3.05	3.04
Surface area (m^2/g)	590	430

Desorption Study: In order to make adsorption process more economical, it is necessary to regenerate the spent sorbent. Attempts were made to desorb As(V) from the loaded sorbents using various molarities of HCl, HNO_3, NaOH and other chemicals. The regeneration cycle was repeated three times. After each cycle the sorbent was washed with distilled water and dried.

Results and Discussion

Effect of Nature of Sorbents

The order of arsenate removal capacities for these chemical adsorbents was found PTS > PLC. This is probably due to several factors such as molecular size, molecular polarity, pH, arsenate reactivity and retentivity, sorbent properties and environmental conditions. The leaching order should be reverse of the sorption order.

Effect of Arsenate Concentration

The sorption of arsenate ions goes on increasing as their concentration decreases from 10 to 0.1 mg/l. The highest removal (minimum leachability) percentages of arsenate ion were 60.5, and 64.8 at initial concentration 0.1 mg/l, pH 6.0, temperature 25°C, flow rate 1.0 ml/min. and 60 cm bed height of adsorbent (Figure 9.1)

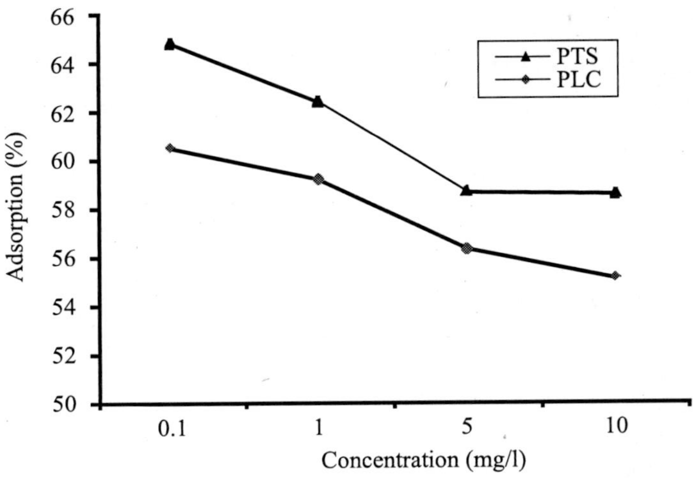

Figure 9.1: *Removal % of Arsenate at different Concentrations and at BH 60 cm, pH 6.0, Temperature 25°C and Flow Rate 1 ml/min*

Effect of Bed Depth

The bed depth also had a significant role in the leachability of arsenate. The study conducted by varying bed depth 15 cm to 75 cm indicated that on increasing bed depth, the leachability of contaminant decreased (Figure 9.2).

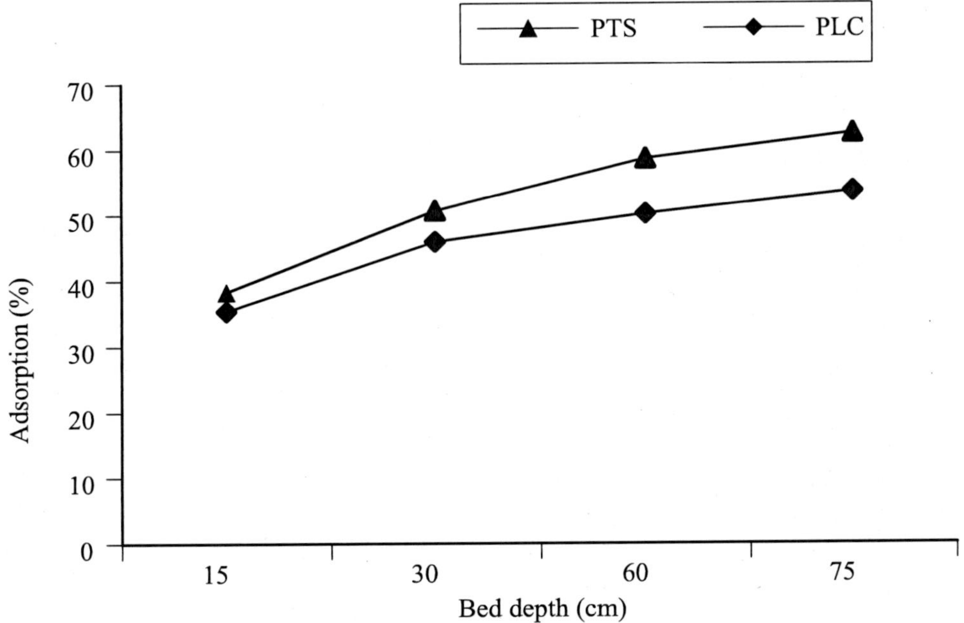

Figure 9.2: *Removal % of Arsenate at different Bed Heights and at Initial Concentration 10 mg/l, pH 6.0, Flow Rate 1.0 ml/min and Temperature 25°C*

Effect of Ph

The effect of pH was studied by changing the pH of the feed from 2 to 8. The solution pH significantly affected the extent of contaminant removal. The

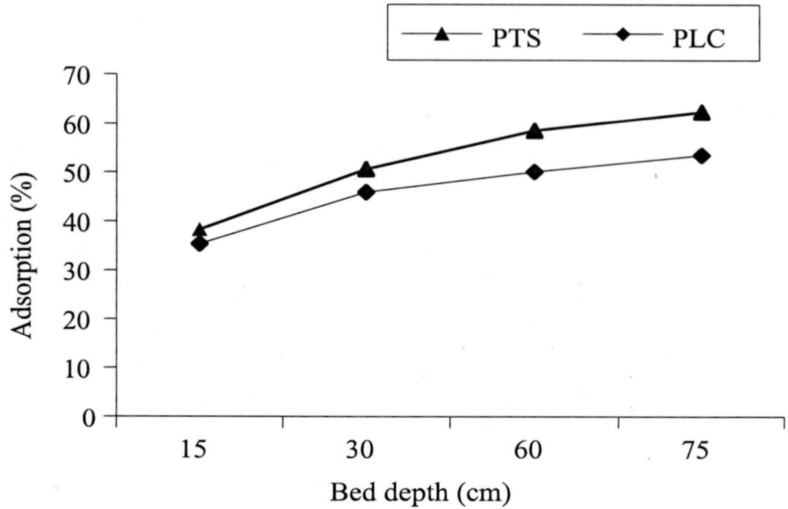

Figure 9.3: *Removal % of Arsenate at different pH Values and at Initial Concentration 10 mg/l, BH 60 cm, Flow Rate 1.0 ml/min and Temperature 25°C*

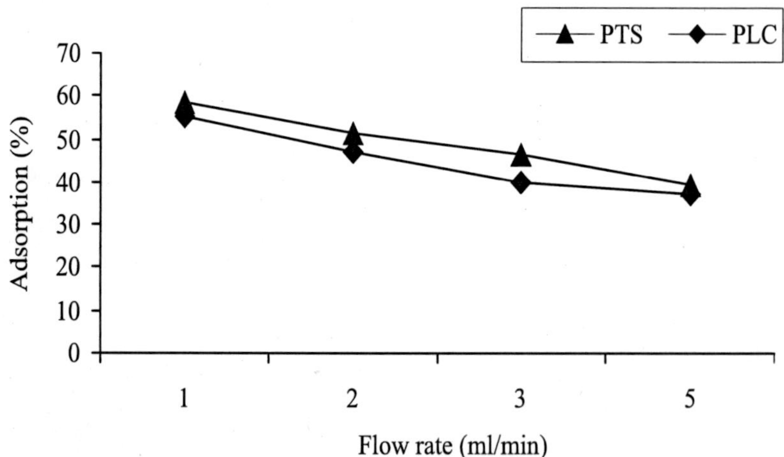

Figure 9.4: *Removal % of Arsenate at different Flow Rates and at Initial Concentration 10 mg/l, BH 60 cm, Temperature 25°C and pH 6*

contaminant removal. The observed results are shown in Figure 9.3. The results showed variation of percent removal at various pH values. Arsenate showed the minimum leaching percent at pH 6.

Effect of Flow Rate

The effect of flow rate was investigated using a column of 60 cm BH and varying the flow rate from 1.0 to 50 ml/min. Leaching increased as the flow rate increased (Figure 9.4).

Desorption Study: Attempts were made to desorb arsenate from the loaded sorbents using various molarities of HCl, HNO_3, NaOH and other chemicals. The regeneration cycle was repeated three times. After each cycle the sorbent was washed with distilled water and dried. Hot and dilute HCl was profitably used to regenerate the spent sorbent.

Pyroaurite

Pyroaurite is a rare clay mineral that comes mostly from the famous mines of **Langban,** Varmland, Sweden, but is also found at a few other localities. It is **hydrated magnesium iron (III) carbonate hydroxide.** It forms platy to tabular crystals. Crystals of pyroaurite can yield flashes of yellow and it is this display that is responsible for its name; which loosely translated means *golden fire*. Pyroaurite is dimorphous with the mineral **sjogrenite.** Dimorphs are minerals that share the same chemistry but have different structures.

Pyroaurite, a magnesium ferric-iron carbonate hydroxide hydrate mineral, was first discovered from Sterling Hill by Paul Desautels. It has not been found at Franklin. Pyroaurite commonly occurs as light yellow to yellowish-brown, hexagonal, platy crystals, up to several mm in width and tabular on [0001]. Pyroaurite crystals also occur “sprinkled” and “impaled” on acicular willemite.

The luster of pyroaurite is vitreous to dull. No physical or optical data exist. A number of small crystals have been examined semi-quantitatively by the writer using microchemical and microprobe methods and were found to be Mg- and Fe^{3+} bearing, with less than 2 wt. % MnO or ZnO.

Pyroaurite is an uncommon mineral locally, occurring on altered vein surfaces, and is associated with willemite, franklinite, calcite, hodgkinsonite, hetaerolite, and other species. In the north orebody, it was found coating haematite and associated with serpentine and calcite. An occurrence on the 1300 level has been reported. These specimens consist of willemite-franklinite ore, with chlorophoenicite, sphalerite, barite, willemite and epitactic sjögrenite as secondary minerals.

Characteristics

- Colour is usually brownish yellow, reddish brown, yellow, white and greenish brown.
- Luster is vitreous to pearly.
- Transparency: Crystals are usually translucent to transparent.
- Crystal system is trigonal, bar 3 2/m.
- Crystal habits include tabular, scaley to platy crystals; also found in fibrous forms.

- Cleavage is perfect in one direction (basal).
- Hardness is 2.5.
- Specific gravity is 2.1 (very light).
- Streak is white.
- Associated minerals includes hydromagnesite, stitchite, calcite, reevesite, sjogrenite, magnetite and lizardite.
- Notable occurrences include the type locality of Langban, Varmland, Sweden; Gulsen Quarry, Kraubath, Styria, Austria; Tunnel Hill Quarry, Tasmania, Australia; Sterling Hill, New Jersey; San Francisco County, California, USA; Half-Grunay, Shetland Islands, Scotland; Rutherglan, Ontario and the Parker Mine, Notre Dame du Laus, Quebec, Canada.
- Best field indicators are crystal habits, cleavage, color and locality.
- Chemistry: $Mg_6Fe_2CO_3(OH)_{16.}4H_2O$
- Class: Carbonate
- Group: Hydrotalcite
- Uses: Only as mineral specimens

A pyroaurite type compound can be synthesized and its properties can be improved for sorption and catalysis purposes (Hansen and Taylor, 1990; Hansen and Taylor, 1991; Hansen and Taylor, 1991; Hansen and Koch, 1994; Hansen et al., 1999; Hansen and Koch, 1995)

General Information

Chemical formula	$Mg_6Fe_2(CO_3)(OH)_{16}\cdot 4(H_2O)$
Composition	Molecular weight = 661.71 g
	Magnesium: 22.04 % Mg 36.55 % MgO
	Iron: 16.88% Fe 24.13% Fe_2O_3
	Hydrogen: 3.66% H 32.67% H_2O
	Carbon: 1.82 % C 6.65% CO_2
	Oxygen: 55.61% O
	100.00% 100.00% = Total oxide
Empirical Formula	$Mg_6Fe^{3+}{}_2(CO_3)(OH)_{16}\cdot 4(H_2O)$
Locality	Langbanshyttan, Sweden
Name origin	From the Greek, pyro and the Latin aurium, "fire" and "golden" because of the gold-like submetallic scales present in its type locality.

Crystallography

Axial ratios:	a:c = 1:7.51857
Cell dimensions:	a = 6.19, c = 46.54, Z = 3; V = 1,544.32 Den(Calc)= 2.13
Crystal system:	Trigonal - hexagonal scalenohedral H-M Symbol (-3 2/m) Space group: R-3m
X Ray diffraction:	By intensity (I/I_o): 7.77(1) 3.89(0.8) 2.62(0.5)

Physical Properties

Cleavage	[0001] Perfect

Color	Brownish yellow, brown white, colorless, greenish, or yellowish white.
Density	2.07
Diaphaniety	Transparent
Fracture	Flexible - Flexible fragments.
Habits	Tabular - Form dimensions are thin in one direction., Fibrous - Crystals made up of fibers.
Hardness	2.5 - Finger Nail
Lustre	Pearly
Streak	white
Optical Properties	
Gladstone-dale	CI meas = 0.025 (Excellent) - where the CI = (1-KPDmeas/KC) CI calc = 0.052 (Good) – where the CI = (1-KPDcalc/KC)KPDcalc = 0.2599,KPDmeas= 0.2674,KC= 0.2742
Optical data	Uniaxial (-), e=1.543, w=1.564, bire=0.0210.

Conclusions

Pyroaurite type sorbents have comparable sorptive capacities to resins. They are easy to be cut, milled or grind and can be used as cartridges for filtering processes. They are highly porous materials, which can be used as ion exchangers, sorbents, filters, ultrafilters, demulsifiers etc. The following applications should be investigated: Purification of drinking water, water softening, purification of radioactive contaminated water and purification of wastewater from hydrometallurgy.

References

Blakely, N.C. (1984), *Behavior of Arsenical Wastes co-disposed with Domestic Solid Wastes.* J. Water Pollution Control, 56(1), 69-75.

Chakarborti, A.K. and Saha, K.C. (1987), *Arsenical Dermatosis from Tube Well Water in West Bengal.* Ind. J. of Medical Research. 85(3), 326-334.

Elson, C.M., Davies, D.H. and Hayas, E.R. (1980), *Removal of Arsenic from contaminated Drinking Water by Chitsin/Chitosan Mixture.* Water Research, 14, 1301-1311.

Ferguson, J.F. and Gavis, J. (1972), *A Review of the Arsenic Cycle in Natural Water.* Water Research, 6, 1259-1274.

Lee, S.H., Lu, Z., Babu, S.V. and Matijevie, E. (2002) *Chemical Mechanical Polishing of Thermal Oxide Films using Silica Particles Coated with Ceria. J. Mater.* Res., 17(10), 2744-2749.

Manju, G.W., Raji, C. and Anirudhan, T.S. (1998), *Evaluation of Coconut Husk Carbon for Removal of As (III) from Water.* Water Research, 32(10), 3062-3070.

Nagarnaik, P.B., Bhole, A.G. and Natarajan, G.S. (2004), *As (III) Removal by Adsorption.* Journal of the Institution of Public Health Engineers, 1, 5-8.

Neeri News (2004), *Arsenic Removal Methods.* 6(1), Compiled by RPBD Division.

Nic, E. and Korte, M.S. (1991), *A Review of As(III) in Groundwater.* Critical Reviews in Environmental Control. 21(1), 1-39.

Nooney, R.I., Thirunavukkarasy, D., Chen, Y., Josephs, R. and Ostafin, E. (2002), *Synthesis of Nanoscale Mesoporous Silica Spheres with Controlled Particle Size*. Chem. Mater.,14, 4721-4728.

Raichur, A.M. and Panvekar, V. (2002), *Removal of As(V) by Adsorption onto mixed Pose Earth Oxides*. Separation Science and Technology, 37(5), 1095-1108.

Saha, J.C., Dixit, A.K., Bandopadhyay, M. and Saha, K.C. (1999), *A Review of Arsenic Poisoning and its Effect on Human Health*. Critical Reviews in Environmental Science and Technologies, 29(3), 281-313.

Seida, Y., Nakano, Y. and Nakamura, Y. (2001), *Rapid Removal of Dilute Lead from Water by Pyroaurite like Compound*. Water Research, 35(10), 2341-2346.

Vaishya, R.C. and Agrawal, I.C. (1993), *Removal of Arsenic(III) from Contaminated Groundwater by Ganga Sand*. IWWA, 21, 249-253.

Wassay, J.A., Haron, M.J., Vchiumi, A. and Tokanaga, S. (1996), *Removal of Arsenite and Arsenate Ions from Aqueous Solution by Basic Yittrium Carbonate*. Water Research, 30, 1143-1148.

OO

10

Prediction of Cookie Quality of Indian Wheat Varieties by Solvent Retention Capacity Profile

Sheweta Barak, Deepak Mudgil and B.S. Khatkar

Abstract

Predictive cookie quality parameters are related to cookie spread ratio i.e. the ratio between the diameter and thickness of cookie. Flours that produce larger diameter and lower height cookies are considered to have better quality. The ability of flour to retain a set of four solvents –water, sodium carbonate, lactic acid and sucrose, produces a flour quality profile for predicting the cookie making performance. The lactic acid SRC is associated with glutenin characteristic, sodium carbonate SRC with levels of damaged starch, sucrose SRC with pentosan and gliadin characteristics; water SRC is influenced by all the flour constituents. Further, AWRC (Alkaline Water Retention Capacity) is a test to select flours of good cookie quality. Excellent cookie-baking flours produce large cookie diameters and low AWRC values. The objective of the study was to evaluate the wheat varieties for their cookie making potential using the solvent retention capacity (SRC) profile and AWRC. The spread ratio of the different wheat varieties ranged from 5.55 to 11.38. The SRC parameters correlated well with the cookie spread ratio. Negative correlations were observed between cookie spread ratio and lactic acid SRC ($r = -0.560$), sucrose SRC ($r = -0.691$), sodium carbonate SRC ($r = -0.553$), water SRC ($r = -0.640$). AWRC values were significantly negatively associated with cookie spread ratio ($r = -0.940$). The regression analysis of the cookie spread ratio on the SRC and AWRC indicated that sucrose SRC, AWRC and water SRC are the strong predictors of the cookie quality. Thus, wheat varieties with lower values of these parameters should be selected for cookie making.
Keywords*: Wheat flour, cookies, solvent retention capacity, alkaline water retention capacity.*

Introduction

The bakery industry is one of the largest organized food industries all over the world and in particular biscuits and cookies are one of the most popular products because of their convenience, ready to eat nature, and long shelf life. A solvent retention capacity (SRC) test method using four solvents (water, sucrose, lactic acid, and sodium carbonate) was published by Approved Method 56-11 (AACC 2000) after Slade and Levine (1994). The percentage of weight increase after centrifugation is the SRC value for each solvent. Generally, lactic acid SRC is associated with glutenin characteristics, sodium carbonate SRC is associated with levels of damaged starch, sucrose SRC is associated with gliadin, and pentosan characteristics, and water SRC is influenced by all of those flour constituents (Slade and Levine 1994; Gaines 2000). All four SRC values differentiate among wheat cultivars across environmental effects with limited genotype × environment interaction (Guttieri et al 2001, 2002). The four SRC values are particularly effective when evaluated as a pattern of four values and correlated with product end-use parameters (Slade and Levine 1994; Gaines 2000).

Desirable cookie and cracker flours have low water-holding capacity (Faridi et al 1994). Lower water absorption by flour provokes more water to be absorbed by sugar, which increments syrup and decreases dough viscosity during baking, that is, dough could spread farther, producing larger diameter cookies (Slade and Levine 1994). Flours with excessive water retention require increased baking times and increased energy costs in bakeries (Guttieri et al 2001). Damaged starch absorbs much more water than undamaged starch, higher levels of damaged starch increased cookie dough stiffness and decreased cookie diameter (Gaines et al 1988). Pentosans are a minor component of wheat flour (2–3%) but they play an important role in dough rheology; pentosans are highly hydrophilic, absorbing as much as 10 times their weight in water (Jelaca and Hlynka 1971). Water-soluble and total pentosan contents have been correlated with smaller cookie diameter and cake volume (Kaldy et al 1991). AWRC is conducted to select flours of good cookie quality. Good cookie flours have low AWRC values and produce cookies with large diameters.

The objective of the study was to evaluate the wheat varieties for their cookie making potential using the solvent retention capacity (SRC) profile and AWRC.

Materials and Methods

Pure wheat cultivars of ten different wheat varieties were selected for the study. The cleaned grains were tempered for 24 h prior to milling. The grains of individual cultivars were milled on a Chopin (Model, CD 1) laboratory mill into flour after tempering. Physico-chemical characterization of the flour of wheat varieties was carried out using standard AACC methods.

Solvent Retention Capacity (SRC) and Alkaline Water Retention Capacity (AWRC) determination of wheat flour fractions.

Solvent Retention Capacity profile (SRC) was obtained according to the AACC 56-11 method. Wheat flour samples (5 g) were suspended in each of 25 g of water, 500 g/l sucrose, 50 g/l sodium carbonate and 50 g/l lactic acid. The samples were hydrated for 20 min and centrifuged at 1000×g for 15 min. Each precipitate obtained was weighed and the SRC for each solvent was calculated using the following equation (Haynes et al., 2009):

$$\% \text{SRC} = \left\{ \left[\left(\frac{\text{tube, stopper, gel wt. - tube, stopper}}{\text{flour weight}} \right) - 1 \right] \times \left(\frac{86}{100 - \text{flour moisture}} \right) \times 100 \right\}$$

Alkaline water retention capacity (AWRC) was determined according to the AACC 56-10 method. Flour (1 g) was suspended in 5 ml of 8.4 g/l $NaHCO_3$, hydrated for 20 min and centrifuged at 1000×g for 15 min at room temperature. The sediment obtained was weighed and the AWRC was calculated. All determinations were made in triplicate.

Preparation of Cookies

Cookies were prepared according to AACC Approved method 10-50D (2000) with slight modifications. The ingredients used were flour (225 g), sugar (130 g), shortening (64 g), dextrose solution (33 ml), sodium bicarbonate (1.6 g), ammonium bicarbonate (0.9g), sodium chloride (2.1 g) and distilled water (16 ml). The dough was sheeted to 10 mm thickness on a sheeting machine and cut with moulder of 60 mm diameter. Baking was performed in baking oven at 205°C for 15 min. The diameter and thickness of six cookies was measured and the average was estimated. The spread ratio was calculated by dividing diameter (mm) with thickness (mm). Cookie preparation was done in triplicate.

Textural Analysis of Cookies

The hardness of cookies was determined by Texture Analyzer. The probe used was Warner –Bratzler blade using 50 kg load cell. The test mode used was force in compression with pre test speed, test speed and post test speed of 1.5, 2.0 and 10.0 mm/sec, respectively.

Statistical Analysis

The Pearson's correlation coefficients for the different quality parameters of the flour fractions of wheat varieties were calculated using SPSS 16.0 software. Multiple linear regression was conducted with cookie spread ratio as the dependent variable.

Results and Discussion

Physicochemical Analysis of Wheat Flours

The results of the flour quality analysis of wheat varieties are presented in Table 10.1. All the flour quality parameters such as moisture, ash, protein, SDS sedimentation volume and damaged starch content varied significantly.

TABLE 10.1

Physicochemical Analysis of the Wheat Flour of different Wheat Varieties

Wheat Variety	*Moisture (%)*	*Ash (%)*	*Protein (%)*	*SDS sed vol (ml)*	*Damaged Starch (%)*
CBW 38	11.90	0.41	11.81	51	8.44
NIAW 917	12.49	0.55	13.03	33	8.85
WH 1021	13.22	0.56	12.68	31	5.11
WH 1025	12.64	0.56	10.62	36	4.68
HUW 234	11.35	0.46	11.37	29	6.76
VL 892	11.32	0.58	11.32	26	4.28
HI 977	12.95	0.55	12.25	52	8.80
HW 2004	11.32	0.54	10.06	47	8.55
MACS 1967	11.06	0.58	12.34	23	8.94
DBW 16	11.35	0.43	12.25	31	8.79

Protein content ranged from 10.06% to 13.03%. The lowest protein content was observed in wheat variety HW 2004 while NIAW 917 showed the highest values, respectively. SDS sedimentation volume ranged from 23ml in variety MACS 1967 to 52 ml in HI 977. The lowest damaged starch content of 4.28 % was reported in variety VL 892 while variety MACS 1967 had the highest damaged starch content of 8.94%.

Solvent Retention Capacity Profile of Different Wheat Varieties and its relationship with Cookie Parameters

Table 10.2 represents the solvent retention capacity profile and the alkaline water retention capacity of the different varieties used in the study. The values for lactic acid SRC ranged from 70.4% in DBW 16 to 90.2 in HI 977; sodium carbonate SRC ranged from 74.13% in DBW 16 to 8.42% in NIAW 917; distilled water SRC ranged from 64.50% in WH 1025 to 76.99% in NIAW 917; sucrose SRC ranged from 84.26% in VL 892 to 97.38% in NIAW 917 while AWRC values ranged from 62.71% in VL 892 to 90.22% in CBW 38.

TABLE 10.2

Solvent Retention Profile and AWRC Values of different Wheat Varieties

Wheat Variety	*LA SRC*	*SC SRC*	*DW SRC*	*SUC SRC*	*AWRC*	*Hardness*	*Spread Ratio*
CBW 38	85.41	84.88	65.89	91.83	90.22	14574.56	6.49
NIAW 917	87.54	88.42	76.99	97.38	89.46	17485.86	5.55
WH 1021	74.15	79.4	68.99	88.08	84.73	5269.50	7.33
WH 1025	71.59	74.38	64.50	86.61	73.39	4870.30	10.31
HUW 234	79.87	78.26	67.47	85.55	82.78	6550.50	8.672
VL 892	70.71	72.47	68.94	84.26	62.71	3845.70	11.37
HI 977	90.2	78.02	75.88	85.61	89.96	8646.96	5.87
HW 2004	81.74	80.54	70.52	87.39	84.62	7784.40	8.61
M A C S 1967	80.32	81.69	69.38	89.15	79.57	10772.45	8.44
DBW 16	70.4	74.13	72.56	92.44	90.21	11940.45	5.67

where DWSRC –distilled water SRC; SCSRC –sodium carbonate SRC; SSSRC- sucrose SRC; LASRC- lactic acid SRC; AWRC- alkaline water retention capacity.

Higher cookie diameter and higher spread ratio are considered as the desirable quality attributes (Yamamoto et al., 1996). The highest spread ratio was observed for VL 892(11.37) while wheat variety NIAW 917 resulted in cookies with poor spread. The hardness values also varied significantly among the different wheat varieties.

In table 3 the correlation analysis showed positive and strong relationships among the different solvent retention capacities and AWRC. Lactic acid SRC was found to be strongly positively correlated with the sodium carbonate SRC while sucrose SRC showed strong correlation with the sodium carbonate SRC. The texture of the cookies measured in terms of hardness correlated positively with the different solvent retention capacities and also AWRC. The spread ratio of the cookies correlated negatively with all the solvent retention capacities and AWRC. Thus, it can be inferred from the data of the study that the spread of the cookies decrease as the values of solvent retention capacity of the wheat variety increase. The results were in concordance with several authors who reported negative correlations between AWRC and spread ratio (Leon et al., 1996, Kisell and Lorenz, 1976, Abbound et al., 1985). Guttieri et al (2001) found that cookie diameter and top grain score is negatively correlated with sodium carbonate, sucrose, and lactic SRC.

TABLE 10.3

Correlation Analysis of the different SRC, AWRC and Cookie Quality Parameters

	LASRC	*SCSRC*	*DWSRC*	*SUCSRC*	*AWRC*	*Hardness*	*SR*
LASRC	1						
SCSRC	0.750*	1					
DWSRC	0.495	0.312	1				
SUCSRC	0.294	0.695*	0.432	1			
AWRC	0.601	0.585	0.468	0.592	1		
Hardness	0.574	0.773**	0.482	0.911**	0.682*	1	
SR	–0.560	–0.553	–0.640*	–0.691*	–0.940**	–0.757*	1

where DWSRC –Distilled Water SRC; SCSRC –Sodium Carbonate SRC; SUCSRC-Sucrose SRC; LASRC- Lactic Acid SRC; AWRC- Alkaline Water Retention Capacity; SR- Spread Ratio.

Prediction of the Cookie making Potential of the Wheat Varieties using the SRC Profile and AWRC

To assess the influence of the factors affecting the cookie spread ratio, multiple regression was used to find equations that could predict the relationship between the solvent retention capacity profile and AWRC and the cookie quality characteristics. Therefore, an equation to predict cookie spread ratio was developed based on multiple regression analysis of the SRC profile and AWRC. The best-fit linear regression model was determined using backward variable elimination. The strong dependence of spread ratio on the distilled water SRC, sucrose SRC and AWRC can be described by the equation:

$$\text{Spread Ratio} = 36.331 - 0.112\,(\text{DWSRC}) - 0.076\,(\text{SUCSRC}) - 0.167\,(\text{AWRC})$$

The above equation had a multiple correlation coefficient of 0.925, which was high enough to predict the spread ratio of the cookies.

Conclusion

The experimental results suggested that the cookie quality is affected by SRC profile and AWRC values of the wheat varieties. It was found that flours with higher values of all the SRC adversely affected the cookie quality. The hardness of the cookies showed a positive association with the solvent retention capacity values indicating that higher values of the SRC make the cookies harder. It can also be stated that flours with higher SRC and AWRC values give cookies with poorer spread. The prediction equation developed by multiple linear regression could accurately predict the cookie spread ratio. Distilled water SRC, sucrose SRC and AWRC were found to be the most important parameters to predict the quality of cookies based on the spread ratio.

References

American Association of Cereal Chemists. 2000. Approved Methods of the AACC, 10th Ed. Methods. The Association: St. Paul, MN.

Faridi, H., Gaines, C., and Finney, P. 1994. *Soft Wheat Quality in the Production of Cookies and Crackers*. Pages 154-168 in: Wheat: Production, Properties, and Quality. W. Bushuk and V. Rasper, eds. Chapman and Hall: Glasgow, Scotland.

Gaines, C.S. 2000. *Collaborative Study of Methods for Solvent Retention Capacity Profiles* (AACC Method 56-11). Cereal Foods World 45:303-306.

Gaines, C., Donelson, J., and Finney, J. 1988. *Effects of Damaged Starch, Chlorine Gas, Flour Particle Size, and Dough Holding Time and Temperature on Cookie Dough Handling Properties and Cookie size*. Cereal Chem. 65:384-389.

Guttieri, M. J., Bowen, D., Gannon, D., O'Brien, K., and Souza, E. 2001. *Solvent Retention Capacities of Irrigated Soft White Spring wheat Flours.* Crop Sci. 41:1054-1061.

Guttieri, M. J., Reuben, M., Lanning, S. P., Talbert, L. E., and Souza, E. J. 2002. *Assessing Environmental Influences on Solvent Retention Capacities of Two Soft White Spring Wheat Cultivars*. Cereal Chem. 79:880-884.

Haynes, L.C., Bettge, A.D. and Slade, L. 2009. *Soft Wheat and Flour Products Methods Review*: Solvent Retention Capacity Equation Correction. Cereal Foods World 54: 174-175.

Jelaca, S., and Hlynka, I. 1971. *Water-binding Capacity of Wheat-flour Crude Pentosans and their relation to Mixing Characteristics of Dough*. Cereal Chem. 48:211-222.

Kaldy, M., Rubenthaler, G., Kereliuk, G., Berhow, M., and Vandercook, C. 1991. *Relationship of selected Flour Constituents to Baking Quality in Soft White Wheat.* Cereal Chem. 65:508-512.

Kisell, L., and Lorenz, K. 1976. *Performance of Triticale Flours in Test for Soft Wheat Quality*. Cereal Chem. 52:638-649.

Leon, A., Rubiolo, O., and Anon, M. 1996. *Use of Triticale in Cookies: Quality Factors.* Cereal Chem. 73:779-784.

Slade, L., and Levine, H. 1994. *Structure-function Relationships of Cookie and Cracker Ingredients*. Pages 23-141 in: The Science of Cookie and Cracker Production. H. Faridi, ed. Chapman and Hall: New York.

Yamamoto, H., Worthington, S. T., Hou, G., and NG, P. 1996. *Rheological Properties and Baking Qualities of Selected Soft Wheats in the United States. Cereal Chem. 73: 215-221.*

OO

Morinda Citrifolia: From Polynesian Uses to Recent Advances

K. Shukla and P. Pant

Abstract

Morinda Citrifolia *(Noni) is an invaluable plant in the coffee family, pantropical in distribution, native to Southeast Asia. Every part of the plant is used. Noni fruits are harvested year round. The fruit is green in colour when raw; yellow when slightly ripe and white when fully ripe with unpleasant odour and strong taste, it resembles the shape of grenade. Direct exposure of Noni to sunlight or to warm temperature after harvest is not a matter of concern; fruits do not get damaged easily and do not require refrigeration. The plant can be grown in infertile, acidic, and alkaline soils. In some settings Noni is considered as weed, as it is difficult to kill, and is one of the first plants to colonize harsh waste areas or lava flows. It exhibits wide range of environmental tolerance including exposure to wind, fire, flooding, and saline conditions. The plant is in use by the Polynesians from over 2000 years. Traditionally leaves and fruit are used for health purpose, to treat acne problems, menstrual cramps, malaria, hernia, tuberculosis, urinary tract ailments, vitamin A deficiency, hypertension, stomach ulcers, and jaundice. The plant has been documented to contain a mixture of over 150 phytochemicals, anthraquinones, Xeronine, organic acids, several vitamins, and minerals like iron, potassium and calcium. The fruit is a good source of carbohydrate, and dietary fibre, but is low in total fats. The fruit can be consumed raw or cooked; seeds of the fruit are edible when roasted. The products of the plant range from whole fruit to juice, fermented juice, capsules, and powder. The products of Noni claim to improve immune system reduce cholesterol and triglycerides in smokers, and treats diabetes. The plant shows analgesic and sedative effects. The plant is used as an anti-cancerous, anti-tumerogenic, antiviral, antibacterial, antifungal, and anti-inflammatory substance.*

Introduction

Morinda citrifolia L. (noni) is an evergreen small tree in the coffee family, rubiaceae, of medicinal importance, traditionally in use from over 2000 years in Polynesia. The plant is native to Southeast Asia and ranges from eastern Polynesia to India. It is believed that the plant was brought to Hawaii by Polynesian ancestors about 2000 years ago when they migrated from South East Asia. Grows in shady forest, open rocky and sandy shore, infertile soil and, it is the first plant to colonize the harsh waste areas and lava flows. The plant is resistant to saline soil, draught, secondary soil, also survives burning. The fruit in spite of its bitter taste is consumed raw with salt by Australian aborigines; it is staple food in Pacific Asian islands. Every part of the plant is used, roots and barks (dyes, medicines), trunks (firewood, tools), and leaves and fruits (food, medicine). The plant has miraculous therapeutic properties like antimicrobial, anti-inflammatory, anticancer, antitumor, antiaeging, anti diabetic, analgesic, sedative, it is also in topical use for wound and burns. Amidst the rising cancer rates, diabetes, asthma problems and the upcoming lifestyle diseases, herbs like Noni are most sought after because of their property of being a link between food and medicine and also the increasing inclination of the modern day people towards naturopathy. The tree has attained significant economical importance as various products of the tree in the form of powder, juice etc made from the fruits and leaves are available in the market.

History of *Morinda Citrifolia*

Morinda citrifolia commonly known as Noni is believed to be second most important plant of the 12 herbal plants brought by the Polynesians. The plant was used by the islanders for treating diabetes, cancer, hypertension, arthritis and pain. There are about 40 different combinations of roots, stem, bark, leaves, flowers and fruit of the plant known and recorded as herbal remedies. The roots of the plants were used to prepare yellow or red dye for tapa clothes and fala (mats) while the fruit was eaten as food. During World War II soldiers based on tropic Polynesian islands were taught by the native people to eat noni fruit to sustain their strength. The fruit was eaten raw or cooked by people of Raratonga, Samoa and Fiji and was staple food for them. Australian Aborigines were fond of the fruit. In Burma, unripe fruits were cooked in curries, while the ripened fruits were consumed raw with salt. Seeds, leaves, bark and root were also consumed by people familiar with the qualities of this unusual noni plant. Noni was considered as a sacred plant in ayurveda and is mentioned in ancient texts as 'ashyuka', which in Sanskrit means longevity. Noni was noted to be a balancing agent, stabilizing the body in perfect health.

Distribution

The plant is pantropical in distribution. Noni is native plant to Southeast Asia (Indonesia) and Australia. The current distribution includes the Indo-Pacific region which includes (Hawai'i, Line Islands, Marquesas, society Islands, Austral, tuamotus, Pitcairn, Cook), Melanesia (Fiji, Vanuatu, New Guinea, New Guinea, New Caledonia, and the Solomon Islands), Western Polynesia (Somoa, Tonga, Niue, 'Uve/Futuna, Rotuma, and Tuvalu) and Micronesia (e.g., Pohnpei, Guam, Chuuk,

Palau, the Marshall Islands, and the Northern Marianas), Indonesia, Australia, and Southeast Asia. The plant has also colonized the open shores of Central and South America (from Mexico to Panama, Venezuela, and Surinam) and on many islands of the West Indies, the Bahamas, Bermuda, the Florida Keys, and parts of Africa. According to McClatchey the seeds can be dispersed by birds and animals also the seeds have the ability to float in oceans.

Uses and Products

Morinda citrifolia is the most important and useful plant in Polynesia. The plant was used by the Polynesians in various combinations for treating menstrual cramps, burns, arthritis, muscle aches, broken bones, deep cuts, sores, wounds, breast cancer, eye problems, diabetes, high blood pressure, cold, influenza. All the parts of the plant are used but the juice of the fruit is most popularly used along with the leaves.

TABLE 11.2

Uses of Different Parts of the Plant

Plant Part	*Use*
ROOT	Yellow pigment (root) and red pigment (bark) used to make dyes, to dye clothing and fabrics, felted clothing (bark-cloth), dyeing batik (Javanese), cotton and wool (Australian aborigines), turbans, carpets, and yarn (Indians). Small pieces of fruit and root kill intestinal parasites. Roots treat stiffness and tetanus and combat arterial tension (Vietnam). Used as febrifuge, tonic and antiseptic.
BARKS	Extracts contain anthraquinones so used to treat ringworm, effective against infectious bacteria strains (*Pseudomonas aeruginosa, Proteus morgaii, Staphylococcus aureus, Bacillus subtilis, Escherichia Coli Salmonella, Shigella).* Skin infections, colds, fevers [10, 11]. Powdered form treats infant diarrhea. Red pigment, used to make dye.
LEAVES	Heated and applied to chest treats coughs, nausea, colic (Malaysia). Leaves juice for arthritis (Philippines). Rheumatism, hypertension, stomach ache, loss of appetite, abdominal swelling, hernia, vitamin A deficiency [12]. Fresh leaves wrapped around fish and meat to impart typical flavor.
STEM	Jaundice and hypertension.
FLOWER	Flower nectarines are very attractive to bees [13].
FRUIT	Antibacterial properties against *Pseudomonas aeruginosa, M pyrogenes, Salmonella montevido, Salmonella typhosa, Salmonella schottmuelleri, Shigella paradys, and E coli* (ripe Noni juice extracts)[13]. Asthma, lumbago, and dysentery (Indo-china). Pounded unripe fruit with salt applied to cuts and broken bones. Ripe fruit is used to draw out pus from an infected boil (Hawaii), juice of over-ripe fruit regulates

Plant Part	*Use*
SEED	menstrual flow & to ease urinary problems in (Malay and Hawaii). Used to make shampoo and to treat head lice. Treats mouth and gum infections, peeling and cracking of toes stomach ulcers [(14)]. Reddish brown, hard seeds covered with protective air sac due to which they can float on water for months. Contains linoleic acid, which is absorbed easily through skin, when applied to skin reduce acne, retain moisture, and acts as anti-inflammatory [(15, 16)]. Roasted seeds are consumed which cleans the body. Used to make insecticide or insect repellent.

Constituents of the Noni Fruit

Extensive research has been carried out on Noni, so as to validate the traditional uses of the plant and in order to discover its constituents which are responsible for the therapeutic effect. The plant is claimed to have over 160 phytochemicals. The fruit juice of *Morinda citrifolia* contains a polysaccharide rich substance Noni-ppt which is reported to have antitumor activity . In order to identify the bioactive constituents the Noni fruit juice was analyzed for chemical and nutritional composition. Thin layer chromatographic methods were employed in order to identify and characterize the bioactive compounds.

The fruit contains a number of phytochemicals like **ligans, oligo and polysaccharides, flavonoids, irridoids, fatty acids, scopoletin, catechin, alkaloids, anthraquinones** (such as rubiadin, nordamnacanthal, morindone, and rubiadin 1-methyl ether, anthraquinones glycoside), **beta-sitosterol, carotene, vitamin A, flavones glycosides, linoleic acid, alizarin, amino acids, acubin, L-asperuloside, caproic acid, caprylic acid, ursolic acid, and Prexeronine**.

TABLE 11.3

List of Pharmacological Activity of Some Active Principles

Active Principles	*Pharmacological Activity*
Oligo- and Polysaccharides	Long-chain sugar molecules, serve probiotic function as dietary fiber, yield short chain fatty acids when fermented by colonic bacteria.
Glycosides	Sugar-phenolic compounds including flavonoids such as rutin & asperulosidic acid, specifically named noni isolates called irridoides and morinidoides have been reported.
Trisaccharide fatty-acid esters, "noniosides"	Results from combination of an alcohol and an acid in noni fruit, noniosides are chemicals that give noni its noxious smell and taste.
Scopoletin	May have antibiotic activities; research is preliminary.
Beta-sitosterol	Plant sterol with potential for anti-cholesterol activity.

Active Principles	*Pharmacological Activity*
Damnacanthal	An anthraquinones having potential as an inhibitor of HIV viral proteins.
Alkaloids	Naturally occurring amines from plants often attribute to causing bitter taste & so may contribute to the foul taste of noni. Some references mention Xeronine or proxeronine as important noni constituents.

Source: Noni Cli. Res. J. 2007, 1(1-2)

Nutritional and Chemical Properties of Noni Juice

The chemical composition of Noni juice depends on the method of extraction, species, and climate, maturity of fruit, storage and on the variability in growing composition. The juice is acidic in nature and contains more than 90% of water of which 50% is in the free form, with very low amount of fat and protein. It is not a good source of basic nutrients as it contains very low amount of protein and fat, with only 38.4% carbohydrates .

TABLE 11.4

Nutritional Composition of Indian Noni Juice.

Constituent	*Percentage*
Moisture (%)	91.6
Crude protein (%)	0.39
Ash (%)	0.46
Crude fat (%)	0.14
Total carbohydrate	3.84
Total dietary fibre	0.72
Energy(Kcal)	154

Of the various micronutrients present in the fruit vitamin A, Calcium, and sodium are present in moderate amounts, and only vitamin C is retained at a high level, 33.6 mg per 100 g of juice.

TABLE 11.5

Main Micronutrients of Nono Pulp Powder

Micronutrient	*Amount* (per 1200 mg)
Vitamin C	9.8 mg
Niacin (Vitamin B3)	0.048 mg
iron	0.02
Potassium	32.0

Volatile Compounds of NONI

According to Farine et al, fifty one detectable volatiles and twenty acids, representing 83% of total volatile from ripe Noni fruit were reported, of which octanoic (58%) and hexanoic acids (19%) and their corresponding methyl and ethyl esters dominated the profile. About thirty two esters were identified of which esters of hexanoic and octanoic acids were the major esters identified.

Scientific Validations on *Morinda Citrifolia*

Dr. Ralph Heinicke: Xeronine System and Cell Regeneration

In a research in Hawaii by a biochemist Dr. Ralph Heinicke a pharmacologically active alkaloid was discovered and named Xeronine. According to Dr. Heinicke Prexeronine is a precursor of Xeronine, and is present in Noni fruit, which can modify the molecular structure of proteins when Prexeronine is acted upon by enzyme Proxerininase in the body and gets converted to Xeronine. Xeronine regulates the rigidity and shape of specific proteins as it can modify the protein's molecular structure.

Anti-Cancerous Activity of Morinda Citrifolia

Annie Hrazumi in 1992 reported anticancer activity from the alcohol-precipitate of Noni fruit juice on Lewis Lung Carcinoma cells in C57 B1/6 mice. The untreated mice injected with active Lewis Lung Carcinoma cells died 9-12 days after injection whereas those treated with Noni juice showed increase in life span. It was concluded that the juice enhances host immune system and acts indirectly. it is not cytotoxic to cell cultures but kills cancer cells by activating cellular immunity. Noni-ppt when combined with sub-optimal doses of standard chemotherapeutic agents such as **Adriamycin (Adria), cisplatin (CDDP), 5-fluorouracil (5-FU), and Vincristine (VCR)** improved the survival time and curative effects proving noni-ppt as supplementary agent in cancer treatment. The S180 ascites tumor growth in mice was inhibited on treatment with noni-polysaccharide **(noni-ppt)** which was characterised as a gum arabic containing heteropolysaccharide composed of sugars glucuronic acid, galactose, rhamnose and arabnose by Eiichi Furusawa and team. Administration of noni-ppt killed the cancer cells by activating peritoneal exudates cell. The peritoneal macrophages when cultured in presence of noni-ppt lead to production of nitrogen oxide and several cytokines (**interleukin-1, tumor necrosis factor, and interleukin-12**). Lymphocytes production was decreased and an increase in interferon-gamma (IFN-r) was observed when lymphocytes were cultured in presence of noni-ppt, IFN-r in turn stimulated macrophages, natural killer cells and cytotoxic cells and lead towards killing of tumor cells. In order to elicit the antitumor potential of noni-ppt the immune cells must be functioning in collectively.

In another study the Tahitian Noni juice was examined for DNA adducts level in peripheral blood lymphocytes of current smokers. The aromatic DNA adducts in peripheral blood lymphocytes was used as a surrogate biomarker for evaluation of environmental carcinogen exposure. About 43.1% reduction in DNA adducts in females compared with 56.1% in males was observed suggesting that drinking Tahitian Noni juice daily may reduce cancer risk by blocking carcinogen-DNA binding or excising DNA adducts from genomic DNA of heavy cigarette smokers. A study on the hypothesis that *Morinda Citrifolia* possesses a cancer preventive effect at the initiation stage of carcinogenesis was carried out by Dr. Wang and the results indicated that Tahitian Noni Juice in drinking water for one week prevented DMBA-DNA (dimethylbenzoanthracene-DNA) adduct formation hence showing a cancer preventive effect in different organs of the mice.

TABLE 11.6

Reduction in Levels of DMBA-DNA Adducts in different Organs of Female SD Rats and Male C57 BL-6 Mice with 10% Tahitian Noni Juice.

Organs	*Reduction in DBMA-DNA Adduct in Female Sd Rats (%)*	*Reduction in DBMA-DNA Adduct Male C57 BL-6 Mice (%)*
Heart	30%	60%
Lung	41%	50%
Liver	42%	70%
Kidney	80%	90%

Anti-Microbial Activity

The plant is known to have anti-bacterial, anti-viral, anti-fungal properties against infectious microbial strains; hence it is widely used to treat infections, wound, bruises, cuts infections, colds, fevers, for making anti-fungal shampoo, for lice treatment and other bacterial-caused health problems. The plant also inhibits the growth of certain bacteria like, *Staphylococcus aureus, Pseudomonas aeruginosa, Proteus morgaii, Bacillus subtilis, Escherichia coli, Helicobacter pylori, Salmonella* and *Shigella*. The anti-microbial property of the plant is because of the presence of phenolic compounds like acubin L-asperuloside, alizarin, scopoletin and other anthraquinones in noni. Scopoletin has shown inhibitory effects on the activity of *Escherichia coli* and *Helicobacter pylori* which is associated with causing ulcers in stomach. According to Rios and Recio in a study regarding comparison of antimicrobial activity of the noni plant with antibiotics resulted in a greater anti-microbial effect by the noni plant proving that the leaf extract of *Morinda citrifolia* can inhibit the growth of the microorganism, and preventing various diseases such as skin infections, diabetes, cancer etc. The alcoholic and petroleum ether extract were evaluated for antimicrobial and anthelmentic activity on *'Pheretima posthuma'*. The alcoholic extract was found to exhibit significant anti-bacterial, antifungal activity comparable to standard drug tetracycline also it produced more significant anthelmentic activity comparable with reference drug Piperazine citrate. The alcoholic extract of the plant may be used for developing some useful drugs against bacterial fungal and anthelmentic action. A study by Locher et al shows that the acetonitrile extract of dried fruit inhibits the growth of Pseudomonas *aeruginosa, Bacillus subtilis, Streptococcus pyrogenes*, and *Escherichia coli.* 1-methoxy-2-formyl-3-hydroxyanthraquinone a compound isolated from the roots of the plant by Umezava and co-workers is known to suppress the cytopathic effect of HIV infected MT-4 cells, without inhibiting cell growth hence showing the anti-viral activity of the plant. The ethanol and hexane extracts of noni exhibit an anti-tubercular effect by inhibiting the growth of *Mycobacterium tuberculosis* by 89-95%. The antimicrobial activity depends on the ripeness and processing of the fruit, the ripened fruit, without drying exhibits a greater anti-microbial activity. *Morinda citrifolia* inhibits the Herpes Simplex Virus activity and it may be due to its principle active constituent's anthraquinones, flavinoid, and alkaloids. The methanolic extract of the stems of the plant exhibits antimicrobial activity against fifteen Gram-positive bacteria, with a inhibition zone of 8-14 mm, twelve Gram-negative bacteria with an inhibition zone of 7-12 mm and thirteen fungi with an inhibition zone of 7-10 mm. It was found to be active against the following microbes. The same author also reported the pesticidal activity of the methanolic extract of the stems against *Aedes aegypti* L. (LC 50- 671.5 ppm).

TABLE 11.7

Anti-Microbial Activity of Methanolic Extract of Stem of the Noni Plant.

Gram-Positive Bacteria	*Gram-Negative Bacteria*	*Fungus*
Staphylococcus AB 188	Salmonella typhi para A	Microsporum canis
Staphylococcus epidermises	Salmonella typhi para B	Tricpyton mentegropyte
Staphylococcus saprophyticus	Shigella flexneri	Microsporangium gypsium
Streptococcus pyrogenes	Proteus mirabilis	Trichopyton tonsurans
Streptococcus fecalis	Proteus vulgaris	Saccharomyces cerevisiae
Bacillus cereus	Escherichia coli	Candida albicans
Bacillus subtilis	Klebsiella pneumoniae	Helementho sporain
Bacillus thuringienisis	Shigella dysenteriae	Aspergillus niger
Micrococcus lysodeikticus	Enterobacter	Aspergillus flavus
Micrococcus luteus	Pseudomonas aeruginosa	Penicillium sp.
Listeria monocytogene		Rhizopus sp.
Corynebacterium xerosis		Fusarium sp.
Corynebacterium hoffmani		
Streptococcus pneumoniae		

Anti-Oxidative Activity

The anti-oxidant properties of methanol and ethyl acetate extracts of the fruit were tested using Ferric Thio-Cyanate method (FTC) and Thio-Barbituric Acid test (TBA) and ethyl acetate extract. The methanol extract of roots of the noni plant exhibited a higher antioxidative activity in comparison with methanol extracts of leaf and fruit. The anti-oxidative activity of ethyl acetate extract of all parts of the plant was higher in comparison to á-tocopherol and BHT. The compounds that contribute to the antioxidative activity of roots may be both polar and non-polar in nature and whereas those of the leaf and fruit may be non-polar in nature. According to a study carried out by Wasina Thani and group, on the anti-oxidative activity of leaf extract, Dichloromethane extract from dried leaves of Thai noni was most potent antioxidant, followed by methanolic extract from fresh leaves.

The anti-oxidative activity of the plant may be due the phenolic compounds, irridoids and ascorbic acid which were detected in noni fruit juice and the anti-oxidative activity was measured in-vitro by Oxygen Radical Absorbance Capacity (ORAC) and 2, 2-diphenyl-1-picrylhydrazyl (DPPH) free radical scavenging methods. Juice from ripe noni fruits contain higher quantities of phenolic compounds, flavonoids, tannins, and scopoletin hence exhibits a greater free radical, superoxide anion radicals and H_2O_2 scavenging activity and higher ACE inhibitory activity. Fresh noni fruit juice is a good source of antioxidant but fermentation, refrigeration, freezing, decrease its free-radical scavenging activity. Noni puree if stored after dehydration exhibits limited reduction of free-radical scavenging activity. Illumination significantly degrades the anti-oxidative characteristics of the noni products by affecting the total phenolic, and ascorbic acid content.

Anti-Inflammatory Activity

In order to study the anti-inflammatory potential of the noni plant carrageenan and bradykinin-induced rat paw model was used. In the rats which were pre-treated either orally or intraperitoneally with noni fruit juice extract, the bradykinin-induced inflammatory response was inhibited and subsided rapidly. The paw edema produced by carrageenan was partially inhibited at low dose and totally suppressed at high dose. Some cyclo-oxygenase, and also in anti-inflammatory activity were selectively inhibited by commercial noni juice. The inhibitory activity of cyclo-oxygenase enzymes (COX-1 and COX-2) enzymes which were involved in breast, colon, and lung cancer was compared with that of commercial traditional non-steroidal inflammatory drugs such as aspirin, Indomethacin® and Celebrex® and the commercial noni juice exhibited a selective inhibition of COX enzyme activity in vitro and a strong anti-inflammatory effect comparable to that of Celebrex® without any side effects.

Analgesic Activity

According to Joseph Betz, Noni fruits possess analgesic and tranquilizing activities. The analgesic and sedative effects of extracts from the noni plant were studied by French team through the writhing and hotplate tests on mice. A significant dose related analgesic activity was observed. A 1600mg/kg noni root extract showed strong analgesic activity similar to the effect of morphine without any addictive or side effects. The findings hence validate the traditional analgesic properties of the plant· In a research by Wang et al by using the "twisted method" where Tahitian noni juice was supplied in drinking water for 10 days. Antimony potassium tartarate used to generate pain by twisting was administered by ip, there was a 82.30%, 74.53% and 64.29% decrease in number of twists in 20%, 10%, and 5% TNJ groups, as compared to control (supplied with drinking water only).

Cardiovascular Activity

In a research by Wang et al it was found that consumption of two ounces of Tahitian noni juice for 30 days improved the lipoprotein profiles in current smokers with 6% decrease in total cholesterol level, 6% and 12% decrease in LDL and triglyceride respectively; while a 16% increase in HDL; 30% and 57% decrease in LDL 3 and 4 respectively was observed. In the electrophoresis map the lipoprotein particles shifted to a larger size; 33% of abnormal type B shifted to normal type A phenotype: a 18% decrease in ratio of cholesterol/HDL, and 22% decrease in that of LDL/HDL, with no effect in the placebo group on total cholesterol, triglyceride, HDL, and lipoprotein. The results suggests that Tahitian noni juice is able to decrease the dangerous sub fractions of LDL (LDL 3and 4) [(40)].

Another study by the same group suggested that Tahitian noni juice may prevent cardiovascular disease in female by improving lipoprotein profiles as a result of improved liver function. The study was carried out in chronic liver injury model induced by CCl_4 in female SD rats which were examined by testing lipoprotein profiles. The rats were supplied with drinking water (control), 10% placebo, and 10% TNJ for four and half months, the rats in placebo and TNJ groups were fed 0.25 ml/kg CCl_4, the rats were then continuously fed with CCl_4 for 12 weeks, after which the rats were supplied with water, placebo, TNJ continuously for another month. The blood sample for lipoprotein profile testing was collected. An increase in total cholesterol, triglyceride, low density lipoprotein, very low density lipoprotein in the CCl_4 group in comparison with control, here the lipoprotein profiles was changed due to CCl_4 induced liver function damage in CCl_4 group. Whereas a reduction in total cholesterol, triglyceride, Low density lipoprotein, very low density lipoprotein was found in the TNJ group.

Conclusion

The scientific studies and validation of the medicinal and therapeutic properties of noni make it a plant of clinical research interest. The plant due to its miraculous properties can be an answer to the modern day lifestyle diseases. The possible studies which can be carried out on the plant may be increasing the fruit- leaf ratio, increasing its strength to withstand the heavy weight during fruiting and winds, the resistive properties to saline water and drought conditions should be considered in order to impart the resistive properties in other plants, the fruit may be made seedless.

References

Wang Mian Ying et al. *Morinda Citrifolia*: Literature Review on Recent Advances in Noni Research (Wang MY *et al*/Acta. Pharmacologica Sinicia 2002 Dec; 23 (1 2): 1127 -1141)

Mathivanan, N., G. Surendiran, K. Srinivasan, E. Sagadevan K. Malarvizhi Intl. J. Noni Res. 2005, 1(1)

Tabrah, F.L., Eveleth, B.M. *Evaluation of the Effectiveness of Ancient Hawaiian Medicine.* Hawaii Med J 1966; 25: 223-30.

Krauss, B. *Plants in Hawaiian Culture.* Honolulu: University of Hawaii Press; 1993. p. 103, p. 252.

Whistler, W. Tongan *Herbal Medicines.* Isle Botanica, Honolulu, Hawaii, 1992. pp. 89-90.

Bruggnecate, J.T. *Native Plants can Heal your Wounds.* Honolulu Star-Bulletin Local News 1992 Feb 2

Abbott, I.A. *The Geographic Origin of the Plants most Commonly used for Medicine by Hawaiians.* J Ethnopharmacol1985; 14: 213-22.

Bushnell, O.A, Fukuda, M., Makinodian, T. *The Antibacterial Properties of Some Plants found in Hawaii.* Pacific Science 1950; 4: 167-83.

Solomon, N. *The Tropical Fruit with 101 Medicinal Uses,* NONI juice. 2nd ed. Woodland Publishing; 1999.

Leister, E. *Isolation, Identification, and Biosynthesis of Anthraquinones on Cell Suspension Cultures of Morinda Citrifolia. Planta Medica.* 1975; 27:214-224.

Maiden, J.H. *Useful Native Plants of Australia including Tasmania.* Sydney: Tuner and Henderson Publisher; 1889.p 45.

Bushnell, O.A., Fukuda, M., Makinodian, T. *The Antibacterial Properties of Some Plants found in Hawaii.* Pacific Science 1950; 4: 167-83.

Scot, C. Nelson, University of Hawai'i at Manoa, College of Tropical Agriculture and Human Resources (CTAHR), Department of Plant & Environmental Protection Sciences (PEPS), Cooperative Extension Service, 79-7381 Mamalahoa Hwy, Kealakekua, HI, USA 96750-7911 USA.

C. Indu Rani, *Noni A Versatile Herbal Plant* Science Tech Enterpreneur feb 2010

West, Brett J.; **Jarakae Jensen, Claude; Westendorf, Johannes** (2008). "*A New Vegetable Oil from Noni (Morinda citrifolia) Seeds*". International Journal of Food Science & Technology 43 (11): 1988–92.

Letawe, C; Boone, M; Pierard, G.E. (1998). "*Digital Image Analysis of the Effect of Topically Applied Linoleic Acid on Acne Microcomedones*". Clinical and Experimental Dermatology 23 (2): 56–8.

Hirazumi, A., Furusawa, E., *An Immunomodulatory Polysaccharide-rich Substance from the Fruit Juice of Morinda Citrifolia* (noni) *with Antitumor Activity*; Phytother Res., 1999 Aug; 13(5):380-7.

Anonymous. Official Methods of Analysis. 6th ed. Washington DC: Association of Official Analytical Chemists; 1995.

Satwadhar, P.N. (2011) *Nutritional Composition and Identification of Some of the Bioactive Components in Morinda Citrifolia Juice*. International Journal of Pharmacy and Pharmaceutical Sciences Vol. 3, Issue 1, 2011.

Farine, J.P., Legal, L., Moreteau, B. and Le Quere, J.L. 1996. *Volatile Components of Ripe Fruits of Morinda Citrifolia and their Effects on Drosophila.* Phytochemistry.41: 433-438.

Wang, M., Kikuzaki, H., Jin, Y., Nakatani, N., Zhu,N., Csiszar, K., Boyd, C.D., Rosen, R. T., Ghai, G. and Ho, C.-T. 2000. *Novel Glycosides from Noni (Morinda citrifolia).* J. Nat. Prod. 63: 1182-1183.

Heinicke, R. *The Pharmacologically Active Ingredient of Noni.* Bulletin of the National Tropical Botanical Garden, 1985.

Hiramatsu, T, Imoto M, Koyano T, Umezawa K. *Induction of Normal Phenotypes in Ras-trans formed Cells by Damnacanthal from Morinda Citrifolia.* Cancer Lett 1993; 73:161-6.

Mian-Ying Wang, Lin Peng, May Nawal Lutfiyya, Eric Henley, Vicki Weidenbacher-Hoper, and Gary Anderson 2009. *Morinda Citrifolia reduces Cancer Risk in Current Smokers by decreasing Aromatic DNA Adducts, Nutrition and Cancer*, 61(5), 634-639.

Wang, MY, Su C. *Cancer Preventive Effect of Morinda Citrifolia* (Noni). Annals of the New York Academy of SCi. 2001 Dec; 952:161-8.

Atkinson, N., 1956. *Antibacterial Substances from Flowering Plants.* Antibacterial Activity of dried Australian Plants by Rapid Direct Plate test. Australian Journal of Experimental Biology 34, 17–26.

Locher, C.P., Burch M.T., Mower, H.F., Berestecky, J. Davis, H., Van Poel, B., *et al. Anti-microbiological Activity and Anti-complement Activity of Extract obtained from selected Hawaiian Medicinal Plants.* J Ethnopharm 1995; 49: 23-32.

Usha, R., Sangeetha Sashidharan and M. Palaniswamy *Antimicrobial Activity of a Rarely known Species* Morinda citrifolia. Ethnobotanical Leaflets 14: 306-11, 2010.

Khuntia Tapas Kumar, Panda, D.S, Nanda U.N., Khuntia, S. *Evaluation of Antibacterial, Antifungal and Anthelmentic Activity of Morinda Citrifolia L.* (Noni). Vol. 2, No. 2, pp 1030-1032, April-June 2010

Y. Chan-Blanco et al. *The Noni Fruit (Morinda citrifolia L.): A Review of Agricultural Research, Nutritional and Therapeutic Properties.* Journal of Food Composition and Analysis 19 (2006) 645–654.

Periyasamy Selvam, Julie M. Breitenbach, Katherine Z. Borysko, John C. Drach. *Studies on Anti-HSV Activity and Cytotoxicity of Morinda citrifolia* L. Noni Leaf. Antiviral Research Volume 90, Issue 2, May 2011, Pages A63-A64.

Bina, S. Siddiqui, Fouzia A. Sattar, Sabira Begum, Tashin Gulzar, and Fayaz Ahmad. *Chemical Constituents from the Stems of Morinda Citrifolia Linn.* Archives of Pharmaceutical Research Vol 30, No 7, 793-798, 2007.

Mohd Zin, Z., Abdul-Hamid, A., Osman, A., *Antioxidative Activity of Extracts from Mengkudo* (Morinda citrifoila L.) Root, Fruit and Leaf. Food Chemistry 78(2002)227-231.

Wasina Thani, Omboon Vallisuta, Pongpan Siripong and Nongluck Ruangwises. *Anti-proliferative and antioxidative activities of Thai noni leaf extract.* SOUTHEAST ASIAN J TROP MED PUBLIC HEALTH. Vol 41 No. 2 March 2010 page no 482.

Shu-Chuan Yang, Tsu-I Chen, Ken-yuon Li and Tsung-Chung Tsai. *Change in Phenolic Compound Content, Reductive Capacity and ACE Inhibitory Activity in Noni Juice during Traditional Fermentation.*

Dussossoy, E., P. Brat, E. Bony, F. Boudard, P. Poucheret, C. Mertz, J. Giamis, A. Michel. *Characterization, Ant-oxidative and Anti-inflammatory Effects of Costa Rican Noni Juice (Morinda Citrifolia* L.). Journal of Ethnopharmacology 133(2011) 108-115.

Yang, J., R, Paulino, S. Janke-Stedronsky, F. Abawi. Free-Radical-Scavanging Activity and Total Phenols of Noni (Morinda citrifolia L.) Juice and Powder in Processing and Storage. Food chemistry 102(2007) 302-308.

Yang, J., R. Gadi, R. Paulino, T. Thomson. *Total Phenolic, Ascorbic Acid and Antioxidant Capacity of Noni (Morinda citrifolia L.) Juice and Powder as affected by Illumination during Storage.*

Marsha-lyn G. Mc koy, Everton A. Thomas & Oswald R. Simon. *Preliminary Investigation of the Anti-inflammatory Properties of an Aqueous extract from Morinda Citrifolia.* Proc. West. Pharmacol.Soc. 45:76-78(2002).

Mian-Ying Wang, Diane Nowicki, Gary Anderson. *The Heart Prevention Study: Improvement of Lipoprotein Profiles in Current Smokers receiving Morinda Citrifolia (noni) Fruit Juice.* Circulation 109, 71-144(2004).

Wang, Mian-Ying, Gary Anderson, Diane Nowicki, Jarakae Jensen. *Cardiovascular Disease Prevention with Morinda Citrifolia (Noni)* by Improving Lipoprotein Profiles in a Chronic Liver Injury Model Induced by Carbon Tetrachloride in Female SD Rats." Circulation Vol. 111, 40-88 (2005). 7,39 793-798, 2007

12

Changes in the Quality Characteristics of Khoa Adulterated with Starch and Refined Wheat Flour

YVS Vynavi, MC Pandey, Rajkumar Ahirwar, PT Harilal and K. Radhakrishna

Abstract

Khoa is a sweetened milk concentrate product which occupies a prominent place in traditional dairy sectors. It is used as a base material in the processing of a variety of sweets and that makes it more prone for fraudulence of adulteration. Different natural adulterants like cheap starch powders and cereal flours are added to increase the weight thereby to fetch more profit. The present study was conducted to find out the changes in the quality characteristics of khoa adulterated with starch and refined wheat flour (Maida) and to establish the control product quality against commercially available khoa. The study was conducted with 10%, 20% and 30% adulteration levels of adulterants and quality attributes of khoa with respect to physico-chemical aspects were studied after adulteration in terms of Hunter color, water activity, proximate composition, and microbial quality. Physico-chemical analysis of commercially available khoa showed significant difference in the color pattern, water activity, moisture, protein, carbohydrate content and microbial quality against control khoa. Khoa adulterated with starch at 10%, 20% and 30% levels showed a decrease in the moisture, ash, acidity and protein with the increase in the concentration of starch. Water activity and Hunter color readings also noted a decrease with the increase of adulterant where as pH increased. Addition of refined wheat flour (maida) at different levels also showed a decrease in the moisture, ash, acidity and protein with the increase in the concentration of starch. Hunter color pattern showed a lighter product with the addition of maida. Microbiologically khoa adulterated with both adulterants were found acceptable in terms of total plate count, coliforms, yeast and molds.

Introduction

India is the largest and fastest growing market for milk and milk products. Market for Indian milk based sweets is developing overseas also. Khoa/mawa as a versatile intermediate base for a wide range of sweets such as burfi and peda, khoa occupies a prominent place in the traditional dairy sector. Khoa is the product obtained by rapid drying milk of cow or buffalo or goat or sheep or milk solids or a combination thereof. It is pale yellow with a tinge of brown, has a sweet taste and a pleasant odor. Khoa contains fairly large amount of muscle building proteins, bone-forming minerals and energy giving fat and lactose. It is also expected to retain most of the fat soluble vitamins A and D, and also fairly large quantities of water soluble B vitamins contained in original milk.

Milk is commonly adulterated with water, starch, or removal of fat, addition of sweetener or color or preservative or addition of milk powder. It was noted that khoa adulteration is rampant. Unscrupulous practices like addition of cheaper ingredients to khoa like starch, neutralizers, refined wheat flour, sugar etc. Considering the above facts, the present investigation was carried out to study the physico-chemical changes and microbiological attributes of commercially available khoa (CK), laboratory processed fresh khoa (LF) and khoa with different adulterants (i.e. starch and refined wheat flour – maida) at varying proportions (i.e.10, 20, and 30%) and to compare them.

Materials and Methods

Samples used for the study are commercial khoa from local market of Mysore (Nandini Dairy), freshly prepared lab khoa sample made from milk with 4.5% fat, 8.5% SNF, Khoa is mainly adulterated to increase the weight, bulkiness and thickness of the sample. During study khoa was adulterated with starch, refined wheat flour at different concentrations. Starch was mainly added to increase the thickness and bulkiness and Refined wheat flour was added to increase the bulkiness and protein content of khoa. The adulterants were selected on the basis of trails with different concentrations based on different concentrations. Based upon different observations, 10, 20 and 30% adulterants were taken for analysis. Beyond 30%, it was clearly detectable by the consumer.

Physical Parameters

Determination of pH: pH was determined **by using** Cyber scan 510 pH meter.

Water activity was expressed as a fraction in pure water $a_w = 1$, in a system without $a_w = 0$. For ideal solution $a_w = M_w$, where M_w is the fraction of water in solution. In milk products, the relationship derives from $a_w = M_w$, especially when milk was highly concentrated. It was determined by water activity (monitor) meter with temperature control, of aqua labs. Sample was filled up to $4/4^{th}$ of a cell and placed into the cabinet. Then knob is twisted in order to take readings. After 15 minutes, it gives readings.

CIE Color Co-ordinates: CIE tristimulus color values X, Y, Z and three dimensional color co-ordinates L^*, a^*, b^* values were measured using D-65 illuminate, with a spectral range 400-700 nm and a spectral resolution of 10 nm (color flex, CFLX-45-2, Hunter lab, Hunter associates laboratory inc., Reston,

VA, USA). The photo sensor was standardized/calibrated using standard black and color tiles and the sample color values were read and recorded using easy match QC software's HL, HAL, Inc Reston, VA, USA) the output values were analyzed for the effect of adulterants on the color changes in paneer sample.

Moisture: The method followed is IS: 10030. 5-10 gms of sample was taken in previously dried and weighed dish. The uncovered dish was heated in a hot air oven maintained at 100 ± 1°C for about 4-5 hrs, cooled in a dessicator and weighed. Process of drying, cooling and weighing at 30 minutes interval was done until the two consecutive weighing is less than 1 mg and lowest weight was recorded and the experiment was carried out in triplicates.

$$\text{Moisture } \% = \frac{(W_2 - W_3)}{(W_2 - W_1)} \times 100$$

where, W_1 = weight of the empty dish, W_2 = weight of the sample + dish before drying, W_3 = weight of the sample + dish after drying.

Ash Determination: This method followed was from AOAC method 900.02 (1997). The sample was subjected to dry ashing by incineration of sample at very high temperature (525°C) in a muffle furnace. 5-10 gms of sample was taken in a tared silica dish and ignited over a low flame to char the organic matter carefully and ignited at 550°C for 4-5 hrs till a grayish color ash is obtained. The silica crucible was cooled in a dessicator and weighed.

$$\text{Total ash\%} = \frac{(W_3 - W_2)}{(W_2 - W_1)} \times 100$$

where, W_1 = weight of empty crucible, W_2 = weight of crucible with the sample before ashing, W_3 = weight of crucible with the sample after ashing.

Acid Insoluble Ash: This method followed was from AOAC method 900.02(1997). The sample after ashing is filtered through ashless filter paper using 25 ml of 10 % HCl. About 10 ml of hot distilled water is used to remove the traces of ash present in the crucible. After filtration, the ashless filter paper is dried in oven and kept in muffle furnace at 550 degree centigrade for 4 to 5 hours. Final weight of crucible with ash is noted down.

$$\text{Acid insoluble ash\%} = \frac{(W_3 - W_2)}{(W_2 - W_1)} \times 100$$

where, W_1 = weight of empty crucible, W_2 = weight of crucible with the sample before ashing, W_4 = weight of crucible with the sample after ashing.

Fat: Fat was determined by Soxhlet extraction method. This method was carried out according to AOAC, 955,04,1997. Empty timble was weighed (W_1). 8-10 gms of sample were transferred into timble and weight was noted down. The difference gives the weight of the sample. A clear, dry flat bottomed standard joint flask (250 ml) with a glass bead was weighed (W_3). Soxhlet extractor was prepared and weighed flask was fixed. The tumble was placed in a extractor and petroleum ether was added in excess. Water condenser was fixed over the extractor (run the extraction set using isomantle with regulation for 14-16 hrs). After 14-16 hrs, flask is disconnected and solvent was allowed for evaporation. Final traces of solvent were removed by heating in oven at 60-70°C for 30 minutes. Flask was cooled and (W_4) was noted down. The experiment was conducted in duplicates.

$$\% \text{ Fat} = \frac{(W_4 - W_3)}{(W_2 - W_1)} \times 100$$

where, W_1 = weight of empty timble, W_2 = weight of the sample, W_3 = weight of flask before extraction, W_4 = weight of flask after extraction .

Free Fatty Acids: Total lipids from the sample were extracted from soxhlet extraction method. After removal of excess solvent, the fat was dried weighed and analyzed for FFA by titrimetric method. The FFA in the given sample is estimated by titrating it against alcoholic KoH in the presence of phenopthalene indicator. The acid number is defined as mg KoH required to neutralize FFA present in I gram of sample. However the FFA present is expressed as milli equivalents.

$$\text{Free fatty acids} = \frac{28.4 \times \text{titre value} \times \text{N of KoH}}{\text{Weight of fat}}$$

where, N = Normality of KOH.

Protein: This method was determined by kjeldhal's method AOAC, 955,04,1997. In kjeldhal's procedure, proteins and other organic digested with sulphuric acid in presence of catalyst. The total organic nitrogen is converted into ammonium sulphate. The digest is distilled with NaOH and titrated against standard hydrochloric acid present. Protein is calculated by multiplying % nitrogen by a factor of 6.25 i.e., percentage nitrogen multiplied by 6.25 = % of protein.

Acidity: This standard method is taken from IS 2785, 1964. 5 ml of potassium hydrogen phalate was carefully pipette out into a clean conical flask and a drop of phenolphthalein indicator was added and titrated against NaOH to a pale pink colour. The normality of NaOH was determined by the formula. N_1V_1 NaOH = N_2V_2 potassium hydrogen phthalate. 10 gms of sample was transferred into a mortar and was ground with small amounts of water, and was transferred into a conical flask were mixed well. 1 ml of phenolphthalein indicator was added and the contents of the flask were titrated against standardized 0.1 N NaOH solution to a pale pink end point.

Titrable acidity as lactic acid

$$\% \text{ by weight} = \frac{9 \times A \times N}{W}$$

A = volume in ml of standard NaOH consumed for titration,
N = normality of standard NaOH solution,
W = weight in grams of sample taken.

Carbohydrate Content: Carbohydrate content = (100 – % moisture + %protein + %fat + %ash)

Microbiological Analysis: Microbiological analysis was conducted in sterile conditions for commercial, lab fresh and adulterated khoas samples. 10 gms of sample was homogenized with 90 ml sterile peptone water. Multiple decimal dilutions were made with the samples diluents. The standard plate count agar on Plate Count Agar (PCA), Yeasts and moulds on Potato Dextrose Agar (PDA) acidified to pH 3.5 using 10% tartaric acid and coliform count on Violet Red Bile agar (VRB) were enumerated.

Results and Discussions

Khoa Properties

Table 12.1 shows proximate analysis of Commercially-available Khoa (CK) and Laboratory-prepared Khoa (LK), which showed that in CK, moisture, acid insoluble ash, fat, free fatty acids, acidity were less than that of LF and the CK did not meet the standard reference values in chemical parameters. In contrast, the protein content and the carbohydrate content of CK were more when compared to the LF and the CK exceeded the limits of the standard reference values. The ash content of the CK declined to a phase which was not at all comparable to the LF which is in 1:15.

TABLE 12.1

Physico-Chemical Analysis of CK and LF

Particulars	*Khoa Samples*	
	CK	*LF*
Moisture (%)	30.25	32.57
Ash (%)	0.06	3.1
Acid insoluble ash (%)	0.02	0.04
pH	6.5	6
Fat (%)	18.53	25.62
Free fatty acids (%)	0.18	0.82
Acidity (%)	0.75	0.93
Protein (%)	20.39	17.87
Total carbohydrate	25.59	20.93
Water activity (a_w)	0.939	0.964

The physical parameters like pH of the CK (6.5) were more when compared to the LF (6), whereas water activity of CK (0.939) was less when compared to the LF (0.964). The colour of khoa sample found using hunter color meter showed that the CK is darker than the LF. It was found from the hunter color values that L, a* and b* values were 76.61, 3.10 and 33.07 in CK while 80.93, 3.47 and 26.43 in LF respectively. When microbiologically tested, CK was detected for coliforms (10^2cfu/g). Standard plate count and yeast and molds were below permissible.

Quality Evaluation of Khoa Adulterated with Starch

The proximate analysis of khoa adulterated with starch (10, 20, and 30%), showed that moisture, ash, acid insoluble ash, acidity, protein decreased with the increase in the concentration of starch (Table 12.2). The carbohydrate content increased with the increasing concentration of starch.

TABLE 12.2

Physico-Chemical Analysis of Khoa Adulterated with Starch

Parameters	*Starch (%)*		
	10	20	30
Moisture (%)	37.32	35.80	34.78
Ash (%)	2.97	2.71	2.54
Acid insoluble ash (%)	0.07	0.06	0.04
pH	6	6	6.5
Fat (%)	5.92	5.71	5.31
Free fatty acids (%)	0.04	0.04	0.06
Acidity (%)	0.75	0.74	0.72
Protein (%)	17.43	15.78	15.68
Total carbohydrate	36.36	40	41.61
Water activity (a_w)	0.999	0.998	0.996

The pH of 10 and 20% adulterated samples were as for 30% sample there is a slight increase in pH. The water activity as well as the colour of the samples decreased with the increase in concentration of adulterant. Hunter color readings in terms of L and a* for 10 to 30% adulteration were decreasing slightly from 80.93 to 79.13, 3.47 to 26.94 respectively while b* values were found to be increased from 26.43 to 26.94 in the same adulteration range. Microbiologically the starch adulterated samples were accepted in terms of coliform count. In contrast the total plate count and the yeast and mold count increased with the increase in concentration of starch from 2×10^2 to 4.3×10^2 cfu/g and $2 \times 10^2$to 8×10^2cfu/g respectively.

Quality Evaluation of Khoa Adulterated with Refined Wheat Flour (Maida)

When khoa was adulterated with maida the moisture, ash, acid insoluble ash, acidity and protein decreased with the addition of maida (Table 12.3). Where as, the fat content, free fatty acids and the protein content increased significantly with the increase in the levels of adulterant.

TABLE 12.3

Physico-Chemical Analysis of Khoa Adulterated with Maida.

Parameters	*Refined wheat flour - Maida (%)*		
	10	20	30
Moisture (%)	36.4	35.9	34.04
Ash (%)	2.79	2.55	2.45
Acid insoluble ash (%)	0.06	0.05	0.04
pH	6	6	6.5
Fat (%)	10.28	10.97	13.49
Free fatty acids (%)	1.08	0.87	0.75
Acidity (%)	0.96	0.9	0.9
Protein (%)	19.57	18.99	18.21
Total carbohydrate	31.59	31.81	32.81
Water activity (a_w)	0.992	0.993	0.995

The pH of 10 and 20% samples were same were as for 30% sample there was a slight increase in pH as similar to the starch adulterated samples where as the water activity increased. The colour of the maida adulterated sample became lighter with increase in concentration. Hunter colour values L and a were found decreasing from 80.93 to 79.13 and 3.47 to 2.95 while b* values increasing from 24.67 to 26.94 for 0 to 30% adulteration.

Microbiologically the maida adulterated samples can also be accepted in terms of coliform count and in terms of total plate count and yeasts and mold count the case was as similar to that of starch adulterated sample. With maximum 30% adulteration of maida showed 2.3×10^2 cfu/g TPC and 1.9×10^2 yeast and molds.

Conclusion

Physico-chemical analysis of commercially available khoa showed significant difference in the color pattern, water activity, moisture, protein, carbohydrate content and microbial quality against control khoa. Khoa adulterated with starch at 10%, 20% and 30% levels showed a decrease in the moisture, ash, acidity and protein with the increase in the concentration of starch. Water activity and Hunter color readings also noted a decrease with the increase of adulterant where as pH increased. Addition of refined wheat flour (maida) at different levels also showed a decrease in the moisture, ash, acidity and protein with the increase in the adulteration of starch. Hunter color pattern showed a lighter product with the addition of maida. Microbiologically khoa adulterated with both adulterants were found acceptable in terms of total plate count, coliforms, yeast and molds.

References

Aneja, R.P. (2002). *Technology of Indian Milk Products.* Dairy Ind, 48(3), 101

AOAC (1990, 1997) *Official Method of Analysis Association of Analytical Chemists* Washington DC.

Balasubramanian, S.C., Lily, G., Mani, G.S., Basu, K.P. (1955) *Nutritive Value of the Proteins of Milk and Some Indigenous Milk Products.* Ind J Med Res, 43(3), 255-64.

Benerjee, A.K., Verma, I.S., Bangchi, B. (1968) *Pilot Plant for Continuous Manufacture of Khoa* Ind Dairy Man 20(1) 84-86.

Das, D.P., Murthy, H.B.N., Swaminathan, M. (1953) *Studied on Nutritive Impairment of Proteins heated with Carbohydrates.* Bulletin of CFTRI, Mysore 2(1) 256-62.

George Borgstrom (1968) *Principles of Food Science* 2(1) 137-138.

Kurein, V. (2007) *World Scenario of Milk Production.* Dairy Ind 6 (1) 7.

Mani, G.S., Lilly, G., Balasubramanian, S.C., Basu, K.P. (1955) *Composition and Nutritive Value of Some Indigenous Milk Products* Ind J Med Res 43(3) 243-48.

Narang, B.D., Kuldip Singh Dhindsa, Kohli, S. (1969) *Physicochemical Studies of Fat Content in Khoa on Storage* Ind J Dairy Sci 22(3) 211.

Rajashekar, K., Kondal Reddy K., Narasimha Reddy K., Sudhakar Reddy K. (2007) *Effect of Processing of Milk into Products on the Residue Levels of Certain Pesticides.* J Food Sci Tech. 44(5): 551-552.

Sethna K., Bhat, J.V. (1949) *Irradiation as a Method for Preserving Khoa* Ind J Dairy Sci 2(1) 12-15.

Sharma, U.P., Zariwala, I.T. (1978) *Survey on Quality of Milk Products in Bombay* J Food Sci Tech 15(3) 118-121.

Srilakshmi (2009). *Milk and Milk Products.* Food science 8(4) 337.

Vivek Sharma, Sumit Arora, Des Raj, Moti Ram and Kamal Kishore (2005) *Incidence of Adulteration in Paneer and Khoa Samples collected from some States of Northern India.* Ind J Dairy Sci. 58(4) 44-45.

Zariwarla (1974). *Survey on Chemical Quality of Khoa.* Ind J Dairy Sci 32(4) 155.

OO

Food Quality Control and Food Safety Management

Kalyan P.Babar and R.S. Gaikwad

Abstract

The International trade in agricultural products is increasingly being dominated by concerns of quality to safeguard human health. For India it is very important that agro-food processing industry improves its functioning and pays attention to hygiene, and SPS standards and also the manufactures/ processors are made aware of the high international standards for quality and safe food. The SPS agreement and the related dispute settlement mechanism of the WTO are an important first step in strengthening the global trade architecture, in bringing greater transparency and orderly conditions to world food trade. International food regulations have emerged as an important source of International trade friction between the developed and the developing countries. The World Trade Organization (WTO) Agreement on SPS (Sanitary and Phytosanitary Measures) provides that these standards and regulations should adopt in food industry for producing quality and safe food.

Introduction

Quality of produce and their products is an objective tool to the producer with respect to the profitability, marketability and consumers acceptability whereas for the consumer, it is a subjective judgment and depends on the position of the recipient in the distribution chain. Quality is the combination of attributes or charactertics of the produce that have significance in determining the degree of acceptability of the product to the consumer. Safe food is an unwritten requirement of all customer specification without any negotiation.

Food quality is an important food manufacturing requirement because the end consumers of food are highly, Vulnerable to any form of contamination that may occur during the manufacturing process. In addition to the quality applied to ingredients, there is also significant need to control the environment where food is produce to ensure that it is hygienic and exposed only to appropriate temperatures. Quality is the main goal of agricultural policy; additional goals are to ensure the food is wholesome, free of pesticides and other contaminants. Other objectives of agricultural policy, such as crop intensive cultivation or introduction of Genetically Modified (GM) crops may not have full consumer support nor be long-term value

Consumer expects safe food and the food industry has a responsibility to meet their expectations. With this objective the quality standards in food industry and the Codex General Principles as a firm foundation for ensuring food safety, is highlighted.

Methods of Quality Control

Traditional Quality Control: The traditional quality control programme was based on establishing effective hygiene control. Control of hygiene was ensured by inspection of facilities to ensure adherence to established and generally accepted Codes of Good Hygiene Practice (GHP) and of Good Manufacturing Practices (GMP). Codes of GHP/GMP are still the basis of food hygiene, However, Codes-although being essential-only provide for the general requirements of the food and the processing of specific food (Marwaha, 2007)

Modern Quality Control: In modern quality control techniques as per the Codex General Principles as a firm foundation for ensuring food safety and their relationship to the development of effective HACCP or equivalent systems. HACCP is a tool to assess hazards and establish control systems that focus on prevention rather than relying mainly on end product testing. HACCP can be applied throughout the food chain from primary production to final consumption and its implementation should be guided by scientific evidence of risks to human health. The application of HACCP systems can aid inspection by regulatory authorities and promote international trade by increasing confidence in food safety. The successful application of HACCP requires a multidisciplinary approach as also the full commitment and involvement of management and the work force. The important parameters to be managed in a company under modern quality control systems are summarized in Table 13.1.

TABLE 13.1

Important Parameters to be Managed in a Company under Modern Quality Control Systems.

Management Concern	*Parameters to be Managed*
Technical management	Intrinsic Quality of Fish (taste, smell and texture), safety, spoilage/freshness, grading, packaging, nutritional, authenticity, shelf life etc.
Managerial concern	Administrative Systems, Customer Relations, Promotion to the product, Delivery commitments, invoicing and payments etc.
Environmental concern	Waste Water Management, Noise Pollution, Odeurs, Pollutants etc.

Source: Development in Food Safety and Quality Systems, Food Hygiene-pp187

Guidelines for Application of HACCP System

Prior to application of HACCP to any sector of the food chain, that sector should be operating according to the Codex General Principles of Food Hygiene. The importance and meaning of quality within HACCP dimensions is the emerging and prevalent concept emphasized for the benefit of supplier and customer. The term quality may vary from market to market as per number of factors where as food safety is one aspect, which is absolute (Dhanakumer, 2004). The important modern method of quality control in food industry involves Good Hygiene Practice (GHP), Good Manufacturing Practices (GMP), Sanitation Standards Operating Practices (SSOP), Quality Control (QC), Quality Assurance (QA), Quality Management (QM), Total Quality Management (TQM) and ISO standards etc.

ISO Standard Requirements

The vast majority of ISO standards are highly specific to a particular product, material or process. The ISO 9000 and ISO 14000 are known as generic management system standards. Over half a million ISO 9000 certificates have been awarded in 161 countries around the world and in 2001 alone over 100 000 certificates were awarded.

ISO Standards	*Description*
ISO 9001	Quality systems – model for quality assurance in design/development, production installation and servicing
ISO 9002	Quality systems – model for quality assurance in production and installation
ISO 9001-2000	It is amendment to ISO 9000 where quality systems can be certified by an external agency and replace the old ISO 9001, 9002 and 9003
ISO 9000	This standard relates to quality management with customer satisfaction and not referred to technical process. It is general quality certificate standard of ISO.
ISO14000	It is primarily concerned with environmental management

Food Safety Standards and Trade: Food safety standards are measures of compliance regulations enacted by the Government to protect the health and safety to their citizens and the environment in which they live. Adopting the promulgation of the Sanitary and Phytosanitary (SPS) Agreement in 1994 as a part to the outcome of the Uruguay round of world trade negotiations, these standards are now popularly known as SPS measures/standards which includes all relevant laws, decrees, regulations, requirements and procedures, product methods and process, testing, inspection, certifications and approval procedures, quarantine treatments. The packaging and labeling requirements directly related to the food safety

Unlike conventional trade policy reforms, SPS regulations can not be implemented simply through legislative declaration. Their effective implementation in the developing countries requires that binding commitments are made to provide adequate financial and technical assistance. In particular, there is a need for a global framework to support national capacity building and improve the design' of international standards (Hoekman, 2002).

SPS Agreement: Indian's Perspective

The SPS Agreement has led to an intensive debate in both the developed and developing countries about its consequences. As tariffs and non-tariff barriers are being lowered due to the WTO commitments, concerns have been expressed that technical measures such as SPS measures are being used to protect the domestic market (Jensen. 2002)

The developing countries are experiencing difficulties in meeting the SPS requirements of the developed countries and concerns have been expressed about the way the agreement is being implemented. The difficulties in exploring under increasingly strict SPS measures are manifold and particularly acute for the developing countries. There is a greater need for expanding the knowledge and understanding of the agreement. The SPS agreement contains a number of instructions that are to be used in achieving its goad. Since its inception, various provision of this agreement by the developing countries as well as India in the WTO. These issues not only involve suggestions relating to the provisions of special and differential treatments (Submission, 1998), technical assistance but also with regard to various other provisions i.e. transparency, harmonization, equivalence etc (I bid, 94)

Food Export Quality Control and Inspection

With the advancement of a food consciousness among consumers, stimulated by the work of the Joint FAO/WHO Codex Alimentarius Commission through its elaboration of food standards, code of hygienic practice and Code of Ethics for International Trade in food, an increasing number of countries have adopted sophisticated food laws and established food control agencies.

Details of food imports released by the United Stated Food and Drug Administration (FDA) indicate that significant quantities of product are at least detained, and at worst rejected, if they fail to meet US food laws. The reasons for detention and rejection of item are summarized in Table 13.2.

TABLE 13.2

Reasons for Detention and Rejection of Food Items between Importing and Exporting Countries.

Detention Reasons	*Rejection Reasons*
Non compliance with labeling requirements	Inability of industries to handle, process, package and transport products to meet the mandatory requirements of importing countries

Detention Reasons	*Rejection Reasons*
Decomposition	Lack of awareness by food exporting countries of the mandatory requirements of importing countries, including certification
Insect and Animal filth and Damage	Lack of adequate export control programme and related agencies in food exporting countries
Use of Prohibited Additives	Lack of communication between food control authorities and agencies in exporting and importing countries
Non compliance with US requirements	
Heavy metal contamination	
Excessive pesticide level	
Excessive mycotoxin level	
Mould infection	
Microbiological contamination	
Swollen and faulty cans	

Current Food Safety Regulations in India with Special Reference to Processed Food Sectors

To give boost to the food industry the need of the hour is to harmonise not only the various food laws but also the agencies. The following are the regulatory/ legislative framework dealing with food safety issues in India for the domestic markets.

Prevention of Food Adulteration (PFA) Act, 1954. The provision of this Act is mandatory and contravention of rules can theoretically lead to both fine and imprisonment. But this happens rarely in practice in India.

Export Quality Control Inspection Act, 1963. This is Act in which the exportable commodities are notified for compulsory pre-shipment inspections. Under these rules quality assurance and monitoring systems manual has been made.

Meat Products Control Order, 1973. This Act covers controlling of meat food products. This act also covers some parts of meat including poultry meat, sea food, etc.

Food Testing Labs. There are 72 food labs under administrative control of the central and state Government as well as local bodies. Four central food labs are established under PFA Act to serve as appellate labs. Samples of food articles taken by food inspectors from state and local levels are tested.

Voluntary Standards

Bureau of Indian Standards: The BIS developed standards for most processed foods that are domestically traded; these standards cover raw material and their quantity parameters, hygiene conditions, packaging and labelling requirements.

Directorate of Marketing and Inspection (DMI), 1937. Under this Act grade standard is prescribed for agriculture and allied commodities like grading and marketing. Grading is voluntary and manufacturers who comply with the standards laid down by DMI are allowed to put Agmark labels on their products.

Ministry of Environment and Forest. This ministry has introduced the Ecomark criteria for certain food items such as edible oil, tea, coffee, baby foods, and processed food and vegetables. This criteria is in accordance with the PFA.

Conclusions

The food industry should see the implementation of SPS, HACCAP and other related standards for food quality and safe food management as an integral part of process of establishing a dynamic business environment in the domestic economy. The scientific and policy research should focus on a strengthening of the linkages between the food and food safety to improve the lives of poor communities by ensuring food security by stronger research and policy frame work in connection with water quality, improving the supply chain and by public awareness.

References

Dhanakumar, V.G. (2007). *Safety Management for Fruits and Vegetables.* Agri. business and food industry 19-24

FAO Agril. Service Bulletin. (1995). *Fruits and Vegetables Processing.* Inter. Book Distributing Co. Lucknow

Food Hygiene (2007). Gene-Tech Books, New Delhi. *Food Safety Regulations Concerns and Trade—A Developing Country Perspective.* Macmillan Indian Ltd. Pub.

Hoekman, B. (2002). *Strengthening of Global Trade Architecture for Development.* The post Doha Agenda, World Trade Review 23-45

Ibid : G/SPS/GEN/94

Jensen, M.F. (2002). *Reviewing the SPS Agreement.* A Developing Country Perspective, CDR working paper

Marwaha, K. (2007). *Development of Food Safety and Quality Systems.* Food Hygiene Gene-Tech Books, New Delhi

Murudkar, M. (2001). *Agreement on Agriculture under the WTO.* Think-Line. A Guna Gaurav Nayas Pub. Nashik

Submission made by India. (1998). G/SPS/GEN/85

Phirke, P.S. (2007). *Post Harvest Engineering of Fruits and Vegetables.* CBS Pub. & Distributors, New Delhi.

WTO, (1998). *The SPS Agreement and Developing Countries* (Geneva, WTO)

WTO, (1999). *Review of the Operation and Implementation of the Agreement on Application of Sanitary and Phytosanitary Measures.*

WTO, (2003). *Committee on Sanitary and Phytosanitary Measures: Specific Concern*

WTO, (2000). G/SPS/W/105

WTO, (2001). G/SPS/19

www.wto.org

OO

14

Dehydration of Vegetables by Application of Radio Frequency Heating

K.P. Babar, P.N. Shastri, R.S. Gaikwad and D.T. Khogare

Abstract

Radio Frequency (RF) energy as emerging drying technologies have drawn much attention from both research community and industry over the past decades. When properly applied RF drying can be energy efficient and of great potential to be used for a wide range of materials. However there are few reviews that cover RF dying technologies. Such a review is timely considering the ever increasing interest of the food industry in adapting energy efficient new drying technologies. The preset investigation was undertaken with an objective to study the dehydration of vegetables by application of Radio Frequency (RF) heating. The study was carried out in the Department of Food Technology, Laxminarayan Institute of Technology; RTM Nagpur University, Nagpur (MS), INDIA; during the year 2007-2008. Observation recorded, shows that dehydration of carrot with initial moisture content 89%, cabbage with 89% and French bean with 90% was carried out in 180 min, 120 min & 180 min respectively. Final moisture content of carrot was recorded as 5%, for cabbage & French bean as 1%. The overall quality of RF dried product was observed to be satisfactory. This may be due to fact that in RF drying water movement through the product facilitated by RF waves, rather than by capillary action, was of happens hot air drying.

Key words: *Radio frequency heating and vegetables*

Introduction

Drying refers to a process in which water is removed from a moist material by using heat as the energy input. The mechanism of drying is a complex phenomenon involving combined heat and mass transfers within a biological food material? Raghavan et al (2005) shows that drying has been reported to account for anywhere from 12% to 20% of the energy consumption in the industrial sector. It is an energy-intensive process because the latent heat has to be supplied to the material to evaporate the moisture. Drying offers a means of preserving foods in a stable and safe condition as it reduces water activity and extends shelf-life much longer than that of fresh foods and agricultural products. A major challenge of drying fresh foods and agricultural products is to reduce the moisture content to a certain low level while maintaining the quality attributes such as color, texture, chemical components and shrinkage. In conventional heating, such as hot air and infrared drying, thermal energy is transferred from material surface to interior due to temperature gradients. These drying processes have low drying rates causing long drying times in the falling rate period of drying. The long drying times at relatively high temperatures often lead to undesirable thermal degradation of the finished products (Mousa and Farid, 2002 and Zhang et al, 2006). Unlike conventional thermal processing in which energy is transferred from a hot medium to a cooler material through convection, conduction and radiation, dielectric heating involves the dissipation of the electromagnetic. Research on RF heating applications in the food industry started in the 1940s (McCormick ,1988 and Anonymous 1993). The first attempts were to use RF energy to cook processed meat, to heat bread and dehydrate vegetables (Moyer and Stotz, 1947 and Kinn, 1947). Thawing of frozen products was the next step on the application of RF energy in 1960s (Jason and Sanders, 1962). Demeczky showed that juices (peach, quince and orange) in bottles moving on a conveyer belt through a RF applicator had better bacteriological and organoleptic qualities than juices treated by conventional thermal methods (Demeczky, 1947. The primary application in the late 1980 s was the post-baking (final drying) of cookies and crackers (Anon, 1987 and Rice, 1993). RF drying methods provide opportunities to shorten drying times and improve the final quality of the dried products. RF energy in industrial drying applications is little known to the public and even within general research communities. Jones reviewed the heating principles, generators and applications of dielectric drying of PVC welding, preheating of moulding polymers and drying of a range of non-metal products. He also reported new developments of dielectric (RF) drying systems Jones (1989). Marra et al. (2009) reviewed the history, basic principle and recent RF heating applications to food. Zhang et al. provided an general review of recent developing in combined microwave drying with other conventional methods (Zhang et al, 2006).

The growth in popularity of convenient foods in many Asian countries has stimulated increasing demand for high-quality dehydrated vegetables and fruits. India can become one of the largest fruit and vegetable exporters in the world and can equally be a large importer given its demographic diversity. This strong footing in agriculture provides a large and varied raw material base for food processing. The Indian food–processing industry is primarily export oriented. India's geographical situation gives it the unique advantage of connectivity to Europe, the Middle East, Japan, Singapore, Thiland, Malaysia and Korea.

Dehydration plays an important role in the preservation of agricultural products like vegetables. They are defined as a process of moisture removal due to simultaneous heat and mass transfer. The most important reasons for the popularity of dried products are longer shelf-life, product diversity as well as substantial volume reduction. This could be expanded further with improvements in product quality and process applications. Vegetables are so common in human diet that a meal without a vegetable is supposed to be incomplete in any part of the world. The preset investigation was undertaken with an objective to study the dehydration of vegetables by application of Radio Frequency (RF) heating.

Materials and Methods

The study was carried out in the Department of Food Technology, Laxminarayan Institute of Technology; RTM Nagpur University, Nagpur (MS), INDIA; during the year 2007-2008. The details of materials used and methods adopted during the present investigation are presented in this chapter.

Materials

The carrot (*Daucus carota*), Cabbage and Spinach were procured from local market of Nagpur city. The vegetables were cleaned and stored properly at room temperature prior to their use in actual experiment.

The carrot, Cabbage and Spinach used in the experiment were obtained from a local market, they were then sorted manually based on their size, washed with Hydrogen peroxide (1%) and then with distilled water and dried on muslin cloth to remove excess water. Carrots were first washed to make sure the carrot tissues were free of mud particles and other extraneous material. The carrots were then peeled using a manual peeler, shredded by shredder and cut into dices of 1-2 mm by using knife.

Chemicals and Glassware

In the present investigation analytical grade chemicals from Himedia, Emerck, and BDH and Glassware from Borosil were used.

Experiment was designed as follows

Dehydration of Carrot, Cabbage and Spinach was carried out by following methods.

Dehydration of Carrot: The previously made carrot dices were directly used for dehydration in RF and carried out dehydration.

Dehydration of Cabbage: The previously pretreated Cabbage were directly used for dehydration in RF and carried out dehydration.

Dehydration of Spinach: The previously pretreated Cabbage were directly used for dehydration in RF and carried out dehydration.

Radio Frequency Drying (RF)

Radio-Frequency dryer, the alignment of molecules of the dielectric materials changes rapidly at the radio frequency field. Since heat is defined as a motion of molecules, increasing the molecular motion generates heat. The electromagnetic energy heats the desired material directly without affecting the surrounding structure or the air within it. The entire material of product is heated uniformly without being dependent on the thermal conductivity of the material Dielectric heating is fast, uniform, energy efficient and clean. RF dryer fabricated by SAMEER with Capacity of 15KW, which is capable of evaporating over 15 Kg of water per hour, was used to dehydrate the Carrot, Cucumber, and red pumpkin at 0.4 to 1 Amp current. The targeted moisture removal kept was below 5% for each product

Moisture Content

About 5g of sample was weighed accurately, spread uniformly into a Petri dish and fire in a hot oven at 105 ± 1°C for 6h. After drying; the covered dish was transferred to the dessicator and weighed soon after it reached the room temperature. The loss in moisture was recorded as the moisture content. After weighing, the procedure of drying, cooling and weighing was repeated until constant weight was obtained.

$$\%\ \text{Moisture} = \frac{\text{Wt. before drying} - \text{Wt. after drying}}{\text{Wt. of sample}} \times 100$$

Dehydration Studies

A known quantity of processed legumes samples was accurately weighed and placed in dryer. Samples were drawn periodically during drying and the moisture content and weight loss was determined by using universal moisture meter.

Calculating Dehydration Ratio

Dehydration ratio was determined by dividing the weight of fresh material taken for dehydration by the weight of dried material. (Ranganna, 1986).

Calculating Rehydration Ratio

Rehydration ratio was calculated as the ratio of the weight of material after and before rehydration. (Ranganna, 1986)

Calculating Coefficient of Rehydration (Ranganna, 1986)

$$\text{Coefficient of Rehydration} = \frac{\text{Initial wt. of sample} \times \left(\begin{array}{l}100 - \text{Moisture}\\ \text{content of sample}\\ \text{before drying}\end{array}\right)}{\left(\begin{array}{l}\text{Wt. of dried sample taken}\\ \text{for rehydration} - \text{Amount of}\\ \text{moisture present in the dried}\\ \text{sample taken for rehydration}\end{array}\right) \times 100}$$

Quality Evaluation of Dehydrated Products

Estimation of B Carotene in Carrot.

The estimation B carotene was done as per standard AOAC method. Extracting samples with 40 ml acetone and 60 ml hexane mixture and measuring absorbance A at 436 nm and calculating with formula gives carotene concentration in mg/lb.

$C = (A \times 454)/(196 \times L \times W)$, where

C = concentration of carotene in mg/lb

L = cell length in cm

W = gm sample/ml final dilution.

Estimation of Total Chlorophyll in Green Vegetables

The chlorophyll in green chili was estimated by measuring optical density of an ether extract of sample at the wavelength 660 nm and 642.5 nm. **(S Ranganna, 1986)**

The total chlorophyll and chlorophyll a and b was calculated by

1. Total Chlorophyll, mg/litre = (7.12 × OD at 660 nm) + (16.8 × OD at 642.5 nm)
2. Chlorophyll a, mg/litre = (9.93 × O D at 660 nm) – (0.777 × OD at 642.5 nm)
3. Chlorophyll b, mg/litre = (17.6 × OD at 642.5 nm) – (2.81 × OD at 660 nm)

Estimation of Acidity in Vegetables

Determine titratable acidity using standard NaOH (0.1) to a faint pink colour with phenolphthalein indicator. (Ranganna-1986).

Result and Discussion

Dehydration of Carrot, Cabbage and Spinach

Drying of fruits, vegetables spices is of great technological interest and has a long tradition as a conservation method. Longer shelf life, substantial weight and volume reduction are the main advantages for its popularity. In the recent years focus has shifted from mere preservation to the preservation of quality, which means improvement in product quality and process application.

In this experimental work we dehydrated carrot, cabbage, French bean by giving pretreatments of steam blanching for 4 min to all vegetables. In addition to we have given 1% Calcium chloride dip for 5 min to carrot to improve texture. And 2% Sodium bicarbonate dip for 5 min for cabbage and French bean for retention of chlorophyll pigment.

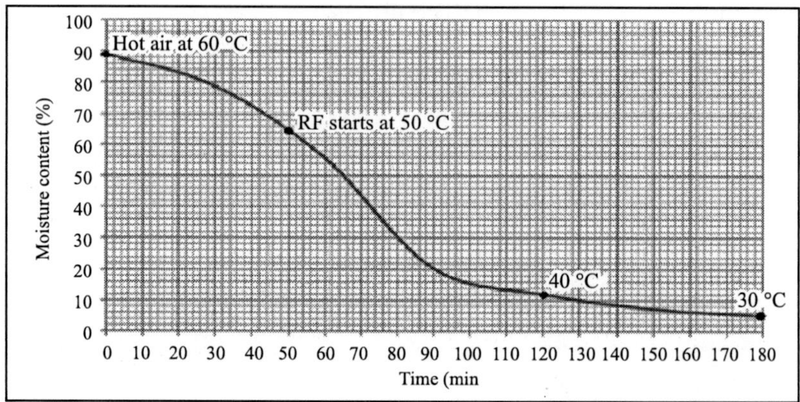

RFD settings-1 hr Hot air at 60ºC and RF for 2 hr at 50 to 30ºC, RF current- 0.8 to 0.5 A

Figure A: *Dehydration of Pretreated Carrot by RF Dehydration assisted with Hot Air.*

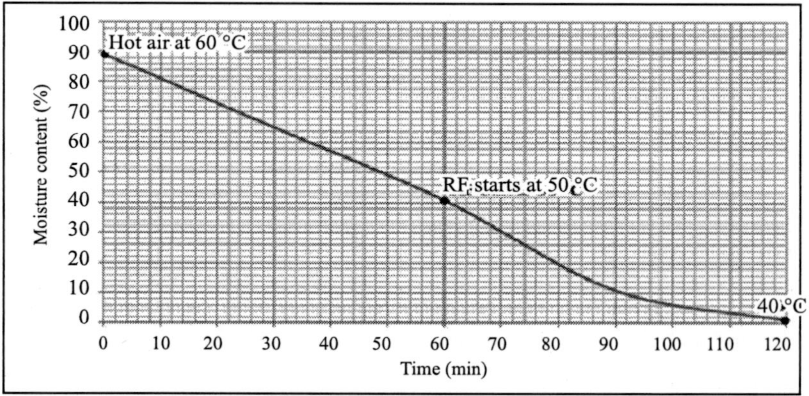

RFD settings-1 hr Hot air at 60ºC and 1 hr RF for at 50 to 40ºC, RF current-0.7 to 0.6 A

Figure B: *Dehydration of Pretreated Cabbage shared by RF Drying assisted with Hot Air.*

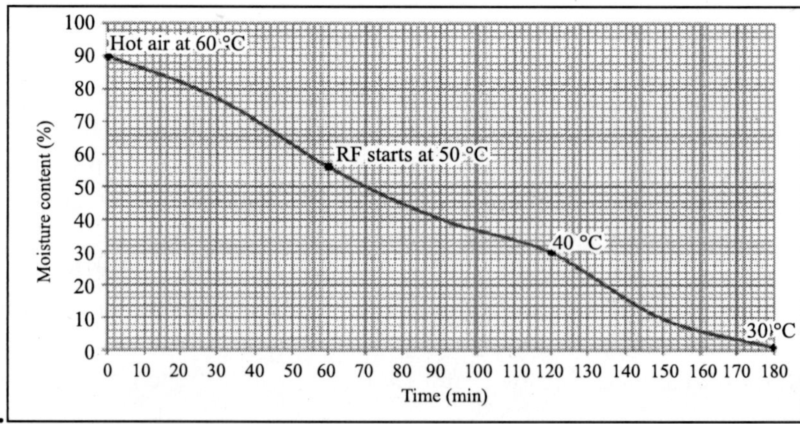

RFD settings- 1 hr Hot air at 60 ºC and 2 hr RF for at 50 to 30 ºC, RF current- 0.5 to 0.4 A

Figure C: *Dehydration of Pretreated French Bean by RF Drying assisted with Hot Air*

Graphs of time Vs moisture content are indicating drying curve characteristic of vegetables.

Observation recorded, shows that dehydration of carrot with initial moisture content 89%, cabbage with 89% and French bean with 90% was carried out in 180 min, 120 min & 180 min respectively. Final moisture content of carrot was recorded as 5%, for cabbage & French bean as 1%.

Rehydration studies were carried out under the ambient temperature to avoid the solid loss. Rehydration was very fast in initial stages and declined thereafter. Rehydration time was same for all vegetables i.e. 120 min. at the end moisture content were achieved 83.7%, 85% & 83% of carrot, cabbage & French bean respectively.

Coefficient of rehydration was calculated by considering the dry weight of vegetables.

Time (min)	*% Moisture*
0	5
30	50.3
60	72.6
90	78.4
120	83.7

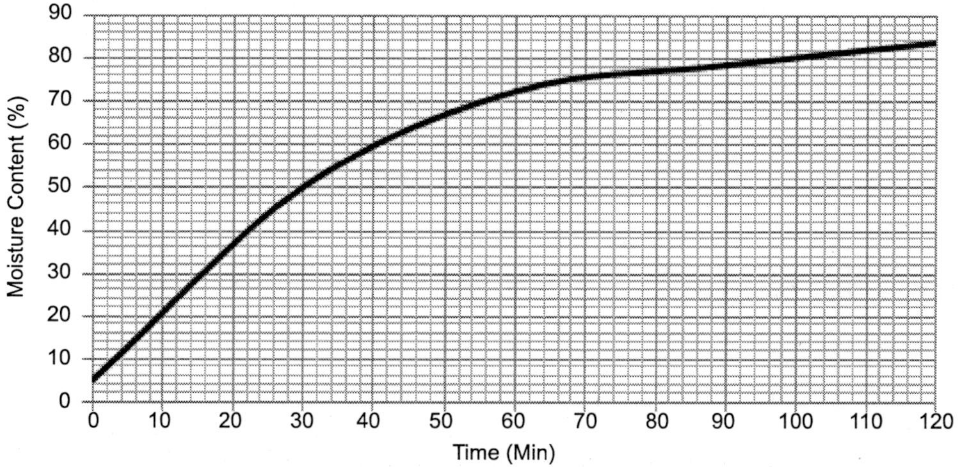

Figure D: *Rehydration Data Curve of RF Dried Carrot.*

Time (min)	*% Moisture*
0	1
30	40.8
60	76.4
90	80
120	85

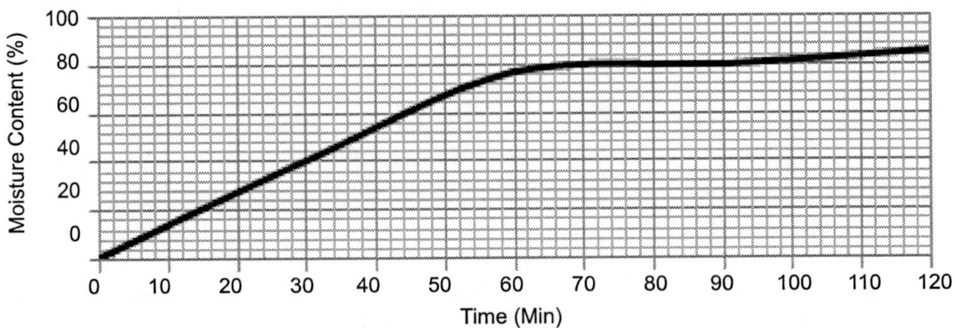

Figure E: *Rehydration Data Curve of RF dried Cabbage.*

Time (min)	*% Moisture*
0	1
30	43.2
60	60
90	75.7
120	84

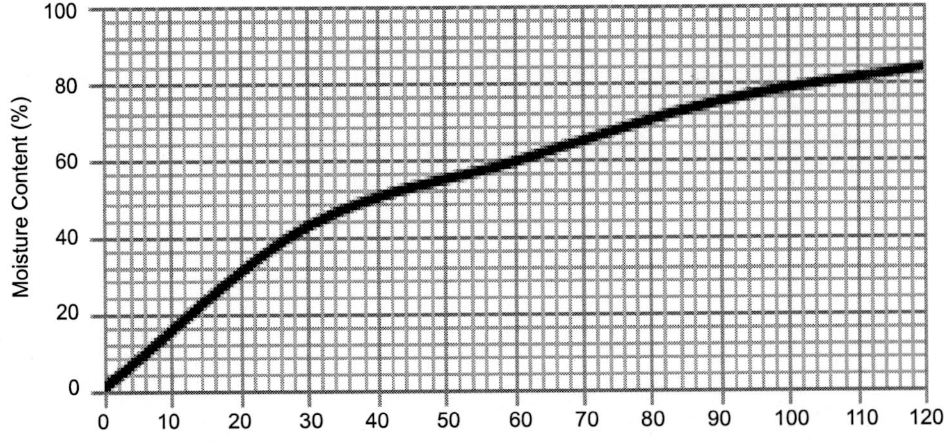

Figure F: *Rehydration Data Curve of RF dried French Bean.*

Carrot and French bean has coefficient of rehydration 0.68 and 0.94 was observed in cabbage. In cabbage coefficient of rehydration was more than carrot and French bean. Cabbage more soft tissues than carrot and French bean.

Percent retention of carotene in dehydrated carrot cubes was 76.08%. in cabbage and French bean observed that percent retention of chlorophyll was 96.55% & 97.68% respectively. Retention of quality parameters like carotene and chlorophyll was more because of pretreatments.

TABLE 14.1

Drying Characteristics of Dehydrated Vegetables

Sr. No.	*Parameter*	*Carrot*	*Cabbage*	*French bean*
1.	Dehydration Time (min)	180	120	180
2.	Dehydration Ratio (D_R)	8.6	13.17	9.54
3.	Rehydration Time (min)	120	120	120
4.	Rehydration Ratio (R_R)	5.78	8.38	6.76
5.	Coefficient of Dehydration	0.68	0.94	0.68

TABLE 14.2

Quality Parameter of Dehydrated Vegetables

Sample	*Parameter*	*Fresh*	*Dehydrated*
French bean	Chlorophyll (mg/100g)*	3.19	3.08
	Acidity on fresh basis	0.03%	0.14%
Cabbage	Chlorophyll (mg/100g)*	3.03	2.96
	Acidity on fresh basis	0.05%	0.20%
Carrot	Carotene (mg/100g)*	3.47	2.64
	Acidity on fresh basis	0.16%	0.64%

*All values are considered as dry weight basis

TABLE 14.3

Coefficient of Rehydration of Vegetables

Sr. No.	*Vegetable*	*Coefficient of Rehydration*
1	Carrot	0.68
2	Cabbage	0.94
3	French bean	0.68

The overall quality of RF dried product was observed to be satisfactory. This may be due to fact that in RF drying water movement through the product facilitated by RF waves, rather than by capillary action, was of happens hot air drying. Thus movement of solids was avoided. RF heats the all parts of product mass simultaneously & evaporates the water at relatively low temperature usually not exceeding 60 p C. Surface discoloration & cracking associated with conventional drying methods are also avoided.

Increases in acidity after dehydration is due to concentration of soluble, however the handling during pretreatments likes 1% $CaCl_2$ dip, 2% Sodium bicarbonate & steam blanching leads to loss of soluble.

Conclusion

Being a new technology the process parameters such as the time temperature schedule for hot air and RF drying, loading density and intensity of RF needs to optimized for the best performance and minimizing the cost. In this present work selected vegetables i.e. Carrot cubes, French bean and Cabbage are dried using radio frequency drying method. Dehydrated vegetables may be used as ingredients for salad mix, soup mix or curries after rehydration. We also used in dressing of many foods after rehydration. In view of hygroscopic nature of dehydrated product, it is recommended that it is packed in laminated pouches preferably under vacuum to retain quality.

References

Raghavan, G.S.V., Rennie, T.J., Sunjka, P.S., Orsat, V., Phaphuangwittayakul, W., Terdtoon, P. (2005) *Overview of New Techniques for Drying Biological Materials with Emphasis on Energy Aspects*. Brazilian Journal of Chemical Engineering, 22(2): 195-201.

Mousa, N. and Farid, M. (2002). *Microwave Vacuum Drying of Banana Slices*. Drying Technology, 20(10): 2055-2066.

Zhang, M., Tang, J., Mujumdar, A.S., Wang, S. (2006). *Trends in Microwave-related Drying of Fruits and Vegetables*. Trends in Food Science & Technology, 17(10): 524-534.

McCormick, R. (1988). *Dielectric Heat Seeks Low Moisture Applications*. Prepared Foods, 162(9): 139-140.

Anonymous. (1993). *Radio Frequency Ovens Increase Productivity and Energy Efficiency*. Prepared Foods, (9): 125.

Moyer, J.C. and Stotz, E. (1947). *The Blanching of Vegetables by Electronics*. Food Technology, 1947; (1): 252-257.

Kinn. (1947). *Basic Theory and Limitations of High Frequency Heating Equipment*. Food Technology and Biotechnology, (1): 161-173.

Jason, A.C. and Sanders, H.R. (1962). *Experiments with Frozen White Fish*. Food Technology, 16(6): 101-112.

Demeczky, M. (1947). *Continuous Pasteurization of Bottled Fruit Juices by High Frequency Energy*. Proceeding of International Congress on Food Science and Technology, 1974: 11-20.

Anon. (1987). *An Array of New Applications are Evolving for Radio Frequency Drying*. Food Eng, 59(5): 180.

Rice, J. (1993) *RF Technology Sharpens Bakery's Competitive Edge*. Food Proc, 6: 18-24.

Jones, P.L. (1989). *Dielectric-assisted Drying and Processing*. Power Engineering Journal, 1989; 3(2): 59-67.

Ranganna, S. (1986). *Handbook of Analysis and Quality Control for Fruits and Vegetable Products*. Tata MaGraw Hill Publishing Company. New Delhi.

Marra, F., Zhang, L., and Lyng, J.G. (2009). *Radio Frequency Treatment of Foods: Review of Recent Advances*. Journal of Food Engineering, 91(4): 497-508.

OO

15

Food Fermentation is Health Friendly: Even if it is Invivo

Dr. (Mrs.) Mukul Sinha

As the wave of healthy lifestyle sweeps across the world, people are looking for a holistic approach to diet and exercise. The health conscious are discerning a lot and view each fruit, vegetable and legume with a new eye. Good health has been discovered in poor men's diet. Although the advantage of roughage/fibre have been known for more than 2000 years, it is only during the past thirty years that fibre is being turned as DIETARY FIBRE with scientific importance. Today the yummy group fat, starch and sugar are suspect whereas a fiber rich diet is being promoted with gusto: Though increasing fibre consumption in diet is difficult because of being less tasty, it is important to develop a product with high fibre ingredients, not only to supply fibre but also provide enhanced functional properties.

Dietary fiber has a long history and several classification systems have been used to classify the dietary fiber. Since most dietary fibers are atleast **partially fermented**, it is suggested that it may be most appropriate to refer them as partially or **poorly fermented** and **well fermented.** Fiber types that are well fermented include pectin, guargum, inulin, gumarabica, oligosaccharides and polydextorose. The partial fermented types include cellulose, wheatbran, cornbran, oat hull fiber and some resistant starches. Generally well fermented fiber are soluble in water, while partially or poorly fermented fiber are insoluble.

Fermentation is the process of bioconversation of organic substances by microorganism and/or enzymes (complex protein) of microbial, plant or animal origin. During fermentation process, microbial growth and metabolism result in the production of a diversity of metabolities. These metabolites include enzymes which are capable of breaking down carbohydrats , protein and lipids present within the substrate and or fermentation medium. While microorganisms are beneficial in most fermentation process, it has additional advantage of being generally regarded as safe (GRAS) for use in clinical nutrition. After more than a century of receiving NOBEL PRiZE by Dr. Metchnikoff for his observation on the use of fermented foods in maintaining health status and well being of people, now it is well recognized that some undigested oligosaccharides act as beneficial **prebiotic** substance for growth of gut microflora in the large intestine of human beings too.

The human being consists of three major interacting and interdependent entities :-

- The **Brain**, the rest of **Body Part** and the **Intestinal Microflora**.

A lack of balance in these three entities of the body is the root cause of illness and therefore treating tissues and organs would not cure the diseases. The gut microflora plays an important role in both human health and diseases.

The **PREBIOTICS** are the nondigestible but **FERMENTABLE** dietary carbohydrate that beneficially affects the host by selectivelyh stimulating the growth and or activity of one or a limited number of bacteria in the colon that can improve host health. Most of the current generations of PREBIOTICS are rather low in molecular weight and they are generally rapidly fermented in the proximal colon. As many chronic gut disorders, such as colon cancer and lucrative colitis originate in the distal colon and progress towards proximal region. It is likely to be as significant health bonus if a prebiotic, saccherolytic fermentation could be achieved in the distal region. The most likely means to achieve this effect is to increase the molecular weight of the PREBIOTIC OLIGOSACCHARIDES. Given the many functions of the intestinal microflora, most attention has been paid to LACTOBACILLI and BIFIDOBACTERA. The reason for this has been mainly intuitive and the evidence of the beneficial effects largely anecdotal.

It is believed that the number of bifidobacterio in elderly people decline significantly after the age of 55 years. PREBIOTICS could certainly be used to target those bifidobacteria that remain in the elderly gut and increase their resistance to invading pathogens.

The major products of the microbial fermentation of nondigestible carbohydrates in the large intestine are short chain fatty acids (SCFAs) mainly acetic, propionic and butyric acids and gases like (H_2, CO_2 and CH_4). These SCFAs are rapidly absorbed from the mucosal surface of the large intestine and used as energy source in many ways. Briefly SCFAs are thought to be efficiently absorbed and utilized by human colonic epithetical cells and to stimulate intestinal motility. Beneficial effects of the intake of nondigestible oligosaccharides on faecal nature and parameters related to constipation have been seen on women and inference drawn that constipation may be improved due to an increase in the osmosis of the increased SCFAs in the intestine by the ingestion of oligosaccharides. The colonic microflora provides the host with a variety of enzymes and the metabolic activity of gut microflora is estimated to b as high as that of livers. The dietary carbohydrates are fermented into SCFAs, gases, protein and amino acids as well as branched chain fatty acids and many more. These substances are absorbed into the blood stream and metabolized in the liver, which show detrimental effects on human health.

By the increased bacterial synthesis of SCFAs in the colon, the absorption of calcium and magnesium also increases as these minerals are absorbed both in small and large intestines. SCFAs increase the passive and active calcium transport in the small intestine or increased transport of calcium from the small intestine to the colon by stimulating mineral absorption by oligosaccharides. The stimulation of absorption seems to be caused by the increased solubility of calcium in the SCFAs rich acidic conditions. Oligosaccharides also recovers the absorption of magnesium and thus suppresses the calcification of the heart and the kidney under the conditions of high phosphorus and calcium dietary concentrates. There are also reports that magnesium iodine and iron metabolism can be improved by dietary oligosaccharides.

India is facing now a days the double burden of undernutrition and overnutrition. While undernutrition is a continuous phenomena since independence of country, the overnutrition problem is strongly related to higher income, low physical activity and the most important is unbalanced intake of nutrients. Therefore the affluent societies are suffering from degenerative diseases like obesity, diabetes, cardiovascular diseases and osteoporosis.

It is in this background that a research project was planned to develop some "Ready to Eat Prebiotic Base Food Mix" which can improve the gut microflora of the diseased subjects and give satiety value to the subjects too. Since prebiotic are oligosaccharides, varieties of pulses used in this study along with vegetable (Bittergourd) and spices (fenugreek leaves and seeds) were considered as source of oligosaccharides.

A total of five RTE mix was developed with four varieties of pulses in common and incorporating fenugreek leaves (ready to eat), seeds, jambu seeds and bittergourd pods differently in each mix base. The diabetic subjects were fed these mix for thirty days equal to a dose of 30/ gm day separately and result obtained were fantastic even in such a short period. The processing included were all home based like soaking, sprouting, drying, roasting and grinding into fine powders. The one kg poly packs were supplied to these subjects with an instruction to eat early in the morning with a glass or two of plain water and salt. A very popular and most commonly used food item used in variety of ways in State of Bihar was selected for innovation purpose known as **SATTU** (Roasted Bengalgram powders).

Reasons for Selection of SATTU

- It is consumed by the almost all the family members of lower, lower middle and upper middle classes as a morning drink.
- It is a proteinceous drink which soothes mucosa of the Gastrointestinal tract after a twelve hour post pranatial period.
- It is Ready To eat Food item which does not require any preparation time.

- If consumed with milk, can provide more nutrient than the costly Horlicks.

Methodology

The methods to innovate this commonly consumed food item was

- Not only Bengalgram but many more varieties of pulses were also included during innovation.
- In addition to variety of pulses some vegetables and spices of medicinal value was also added.
- The method of simply roasting was supplemented by many more home based processing methods to increase palatability and functionality.

Presumed Mechanisms of Action

Since whole pulses and vegetables have undigestible but fermentable oligosaccharides, which when passes through distal end of colon without being affected by digestive enzymes will be attacked by the gut microfloera to produce many health friendly items as mentioned in the reviews cited earlier. Further pulses have low glycemic index, fenugreek is rich in an alkaloid named Triginolline which stimulate hormone insulin to utilize blood sugar, bittrergourd has a popypeptide – P which acts as plant insulin. Therefore also functions as depressor of blood sugar level. The hypoglycaemiic agent CHARANTIN increase glucose uptake and converting the same into the glycogen in the liver.

So, if all these ingredients are used to develop SATTU in an innovative way, this can not only help diabetic people to control their blood sugar level but will have a preventive effect on susceptible population too.

Conclusion

The human gastrointestinal tract forms a variable ecosystem with a wide range of different conditions. The mucosal surface provides a large area for the adherence to and microbial colonization of the small intestine. The pH of empty stomach very effectively eliminates most microbial. However the buffering effect of foods allow the survival of salivary bacteria and microbes present in the ingested foodstuffs in the duodenum. The gastric content is nutrilized by pancreatic secretion, thus further enhancing microbial survival. On the other hand the presence of bile and pancreatic enzyme form an extra stress for the bacteria.

One of the most positive functions of intestinal bacteria is the prevention of bacterial and viral infections by preventing pathogens establishing themselves or colonizing the intestinal tract. Further, they are involved in the metabolism of bile acids and xenobiotics. An important phenomenon is the formation of butyrate and other short chain fatty acids as products of **Carbohydrates Fermentation** by the strict anaerobes in the gut. Butyric acid appears to have a range of physiological functions in the gut epithelium both as nutrients and as a controller of gene expression.

With the same intension the development of a large and potential market for fibre rich products with healthy outcomes have flourished in India.

Although there have been great achievements in this research field, further investigation are needed for designing "Ready to eat Foods" that consider the precise functionality of dietary fiber from both technological and physiological point of view.

References

Electronics Forum on Biotechnology in Food and Agriculture: Conference 11 of FAO on Biotechnology Application in Food Processing: Can Developing Countries Benefit (2004):1-10.

Nutrition News, Vol-8, No. 3, 1987.

Nutrition News, Vol.-27, No. 3, 2006

Food & Nutrition, News, ANGRAU, Vol. 1, No. 1, 2010..

Marcel, B., Roberfroid (2000) *Prebiotics and Probiotics: Are They Functional Foods*. Am. J. Clinical Nutrition (Suppl.):16823-7S.

Poonia, Amrita and Dabourm R.S. (2009): *Functional Foods-New Opportunities and Challenges in Production*, IFI, Vol-28, No. 3, 53-61.

Sharma, B.R., Rana V. and Naresh L., (2006): *An Overview on Dietary Fiber*, IFI, Vol. 25, No. 5:39-46.

Shriwastava Swati and Goya G.K. (2007). *Therapeutic Benefits of Pro and Pre Biotics: A Review*, IFI, Vol. 26, No. 2:41-49.

Nair, M, Babu. et al (2006) *SASNET-Fermented Foods: An Initiative for Meeting the Challenges of Poverty and Hunger*. IFI, Vol-25, No. 5:13-15.

Madar & Odes, 1990s Lefebire and GThebaunctin), 2002

Sampath, S., and Kondal, Reddy. (2005). *Secreating of Cereals and Pulses for Oligosaccharides Content Food and Nutrtion News, ANGRAU, Vol. 1:5-6*

16

Nutritional Status of Berseem in India

Siddharth Priyadarshi, Anubha Shukla, Prity Pant

Abstract

Barseem (Trifolium alexandrium) also known as Egyptian clover is an animal fodder. It is a winter forage legume in India, Pakistan, Turkey and Egypt. Barseem is the most important winter season legume cultivated in an area of around 2 million hectares in India, particularly, central, north, north-west and north-eastern parts. Sowing is to be done in mid October. The crop can be harvested in 45 days after sowing and subsequent cuts can be taken after every 30 days. It has good forage quality (20% Crude protein), digestibility (up to 65%) and high palatability for cattle. In fact, it is known as milk multiplier of cattle's. Being a leguminous crop it fixes atmospheric nitrogen and thus is also good for soil fertility. Around seventy per cent of the adolescent girls of India are anemic, as they require 15 % Additional iron and folic acid. In the under privilege group like orphans cheap and effective methods to treat anemia are a challenge. Leaf Concentrate (LC) is a dark green extract from Berseem which is extremely rich in malnutrients especially iron and folic acid. LC is known to improve the general health and particularly hemoglobin (Hb) levels. Plant foods like leaves contain almost all of the mineral and organic nutrients establish as essential for human nutrition, as well as a number of unique organic phytochemicals that have link to the promotion of good health. Nutritional status of subjects will assess using anthropometry, dietary, clinical assessment and hemoglobin estimation.

Introduction

Adolescence is the process whereby an individual makes the gradual transition from early days to middle age, very habitually it is equated with youth and cycle of physical changes culminate in reproductive maturity. Adolescence is a period of great physical and physiological stress especially in girls as there is onset of menstruation. At menarche most girls become anemic.

Iron deficiency anemia is the commonest cause of nutritional anemia in India, which is indicate by reduction in concentration of hemoglobin in the peripheral blood below the accepted normal levels for the specific age group and sex of individuals. Several studies from various part of the country indicate that anemia in 10-20 year old group is ranging between 50-70%. In developed world prevalence is 10%, in developing world it is 40% and in totality it is 35%. Compare to these figures India has significantly higher number of anemic. Running anemia is one of the biggest challenges in India. Probing for original and cheaper options is the need of the hour. Leaf Concentrate (LC) is not a new food but simply a novel and better way of offering greens humans. It is a product made by fractionating green vegetation and it is perhaps the richest untouched food resource on earth. Leaves commonly used to extract LC are Lucerne (*Medicago Sativa L*), Berseem (*Trifolium Alexandrium L*), Green Leafy Vegetables like radish, turnip, carrot, beet root tops and even obnoxious weeds like congress grasses could be used for extracting LC.

LC is highly healthy food. It is the richest source of iron folic acid. As little as 3 gm LC supplies the whole day requirement of beta-carotene and about 15 gm of LC meets over 75% of other nutritional need in the diet and therefore LC scores over other alternative food sources. The feeding trials with LC supple-mentation have been carried by various Nutritionists and food scientists and positive results have been reported on growth and hemoglobin levels, improving nitrogen retention, increase in serum protein and spontaneous revision of anemia.

Social and Nutritional Status in India

Real India lives in villages, where electricity, clean water, sanitation, education, medical facilities is either no or less are available. Due to male dominating society, girl foetus is to be killed and if they survive, nutrition diet is not provided to them. India has > 50% under malnourished mothers and maternal anemia in >85% (Agarwal et al. IJMR, 2006). 30% births are low birth weight. 47% children <3 years (especially girls) are malnourished. So it becomes very important to empower women by providing education and ways to earn money. It can be done by giving appropriate training and demonstrating low cost and easy to understand technology to develop a new product from an easily available raw material, the Barseem fodder crop.

Conception

The protein content of common Indian leafy vegetables (100 gm. Eatable portion) is Amaranth-2.5g, Mustard-2.7g, Turnip- 1.5g, Broccoli-1.8g, Cauliflower leaves-2.0g, drum sticks (dried leaves)-29g. However the micronutrient content is similar in these leaves. Lucerne (Alfalfa), green fodder has been in use by homeopaths as medicine (tonics). Barseem is also a green fodder. Its fresh leaves contain 18-23% protein and is very rich in minerals and vitamins. Barseem is successfully used for animals but not at all for humans. The concept of utilizing Barseem in an acceptable form is worth drying, using available non sophisticated technologies.

Design

- Optimization for the extraction of Barseem.
- Study Physical-Chemical and microbiological characteristics of the extract.
- To develop process parameters for the preparation of powder from the extract.
- To study the Physical- Chemical and microbiological characteristics and shelf life of powder.
- To study the nutritional benefits on rural people.
- To give demonstration and training to rural people to utilize this method for small scale industry.

Plant Leaves with Immunomodulating Properties

So far the known leaves with such properties are:

- Tea leaves for decreased pro-inflammatory cytokine effect- (Neyestani Intl J food Sci & Nutr 2008)
- Azadirachta Indica (neem leaves) for enhancing TH -1 immune response (mandal Ghosh et al 2007; Cancer Immunity).

Conclusion

Thus it can be concluded that leaf concentrate is a novel product which can be used for supplementation in various conditions like anemia. LC can bring about significant changes in the health of an individual suffering from common deficiency diseases arising from iron, folic acid and vitamin A. We are developing one hypothesis in which the hemoglobin levels will estimate at baseline, mid-intervention and after intervention. After using Berseem period of 3 months significant increase in hemoglobin levels will observe.

References

Nimali Singh, Reshma Boolchandani et.al, Efficacy of Iron Rich Leaf Concentrate on the Haemoglobin Level of Adolescent Girls (13 to 18 years) in an Orphanage of Jaipur city.

Sachdev, H.P.S. and Choudhary, P. (2004). Nutrition in Children: Developing Country Concern, First Edition, B.I. Publications Pvt. Ltd. New Delhi: 217-236; 272, 492.

Palta, A and Gurwara, N. (2003). *Haemoglobin and Cardiovascular Efficiency of Adolescent Girls*, Ind. J. Nutr. Dietet,; 40: 327- 332.

P. Dewan, K N Agarwal Berseem (*Trifolium alexandrinum*) *Leaves in Diet as Immuno-nutrient; Cytokine and T-cell Subpopulation Response in Malnutrition.* Department of Pediatrics, University College Medical Sciences, Delhi INDIA.

Mehta, M.B. and Dodd, N.S. (2004). *Effect of Different Levels of Iron Supplementation on Maternal Iron Status and Pregnancy Outcome.* Ind. J. Nutr. Dietet., 41: 467.

❍❍

17

Quality Changes in Milk Adulterated with Detergent, Urea and Neutralizers

MC Pandey, Rajkumar Ahirwar, PT Harilal, K Radhakrishna and AS Bawa

Abstract

Milk adulteration is a common practice found throughout India especially during festival seasons. The chief objective of adulteration of milk is to derive extra profit sometimes (by adding water or extraction of fat as it is a most valuable constituent in milk). Besides this, milk is also adulterated with synthetic milk, a white milk-like liquid made out of urea, detergent and vegetable fat. In the present study, quality changes in milk adulterated with different adulterants like urea, detergent and neutralizers were carried out. Adulterants were added in different concentrations and their effect on the physico-chemical and microbiological quality characteristics were studied. Addition of detergents at 0.1, 0.3, 0.5, 0.7, and 0.9% concentration showed an increase in the Refractive Index (RI) from the control value of 1.344 to 1.345, 1.346 1.347, 1.348 and 1.350 respectively. Specific gravity also increased from 1.030 to 1.0305, 1.0315, 1.0320 and 1.0325 respectively. Same trend was observed with total solids, total ash and pH; Whereas, moisture, fat lactose content decreased. The milk adulterated with urea at 0.5, 1, 3, 5, and 7% had a pH of 6.68, 6.79, 7.21, 7.64 and 8.35% respectively showing a significant increase. Increase was also observed with total solids, viscosity, RI and specific gravity. Moisture, fat and lactose content decreased with the increasing concentration of urea. Addition of neutralizer showed an increase in the pH, solids not fat (SNF), RI, specific gravity and total solids at 0.1, 0.5, 1 and 2% levels. Microbial growth in the adulterated milk did not show much variation however slight increase in the colony count was observed with total plate count, coliforms, yeast and molds in all adulterated samples. Based on these studies sample field kits have been developed to detect these adulteration in milk.

***Keywords**: Milk , Urea , Detergent , Neutralizer, Microbial analysis and Hunter color.*

Introduction

Milk is an essential nutritional food for infants, adults as well as aged (Potter, 2005). Milk is known to contain over a hundred biochemical compounds of which the majority has substantial nutritional value. The dry weight content of milk is around 13%, including milk fat (3.5-5%), protein (2.5-5%), carbohydrate (mainly lactose 4.6-4.8%) minerals, vitamins, enzymes, hormones pigments. Milk protein covers all essential amino acids which the human body needs on a daily basis for it does not produce them itself (Omkar and T. Rai, 2010).

Milk Adulteration

Milk is mainly adulterated with water in India by vendors. Addition of adulterants like water, starch, salt, pulverized soap, detergents, urea, skim milk powder and preservatives like formalin, hydrogen peroxide etc., in the milk deteriorate its overall quality. They are used to prepare synthetic milk also (Rhodes, 2002). The most common method of milk adulteration is to increase the volume by addition of water or extraction of fat as it is most valuable constituent of milk. As addition of water changes the appearance of milk and lowers its specific gravity. Sometimes to mask the appearance and to increase the specific gravity after the addition of water, some soluble solid substances like sugar or starch is added. In detecting adulteration in milk, the percent of the main constituents is considered with reference to the average standards or legal standards to those constituents. The common method of adulterating milk is by dilution with water. This destroys the natural yellowish white color this can be compensated by adding urea to get milky white. The color change in milk can be easily estimated by hunter calorimeter.

The present study was conducted to find out the changes in the quality characteristics of milk adulterated with urea, detergent and neutralizer to establish the milk quality against control. The study was conducted with 0.1%, 0.3%, 0.5%, 0.7%, and 0.9% detergent, 0.1%, 0.5%, 1%, and 2% neutralizer and 0.5%,0 1%, 3%, 5%, and 7% urea respectively. The quality attributes of milk with respect to physico-chemical aspects were studied after adulteration.

Materials and Methods

Fresh homogenized milk was obtained from M/s Nandini dairy, Mysore for conducting different experiments. Percent adulteration of milk was done by mixing appropriately weighed adulterant (as per % weight) and making up the volume using milk to 100 ml.

To find quality changes in colour the CIE tristimulus colour values x, y, z three dimensional colour co-ordinates L^*, a*, b* values were measured using D-65 Illuminant with a spectral range of 400-700 nm a spectral resolution of 10nm (Colour Flex, CFLX-45-2, Hunter lab, Hunter Associates laboratory Inc., Reston, VA, USA). The photo sensor was standardized/calibrated using standard black and white colour tiles and the sample colour values were read / recorded using Easy Match QC Software. The output values were analyzed for the effect of adulterants on the colour changes in milk.

Refractive index (η), was determined using Abbe's refractometer (Advance Company, R-8, No. 23017), pH determinations were made using digital pH meter Cyber scan 510. Moisture was found using Hot Air Oven method (AOAC. 1990. Volume 2) and ash content by AOAC method 900 02 (1997). Similarly acidity was estimated using IS method 1479 (II) 1961, fat determination was conducted by the Gerber method, IS: 1224, (Bureau of Indian standards, New –Delhi), total solids by IS: 1479 (Part II) – 1961, casein by IS: 1479 (Part II) – 1961 and specific gravity by conventional method (using lactometer).

Microbiological Analysis

Studies at different percentage (%) levels of adulterated milk with urea, neutralizers and detergent were carried out to establish the microbial quality as per APHA (1992) procedure for Standard Plate Count (SPC), coli form, yeast and molds.

Result and Discussion

Adulterated Milk with Detergent

Table 17.1 indicates the changes in the quality parameter in milk adulterated with detergent. The moisture content of the raw milk was 88.35%. Milk adulterated with 0.1% of detergent showed 86.74% of moisture. Similarly with 0.3, 0.5, 0.7 and 0.9%, the readings were adulterations obtained as 86.10, 85.03, 84.86 and 83.61% respectively. Therefore with the increase is the level of detergent the moisture content of the milk was decreased and the reduction in moisture was due to dissolved detergent. The total solid content of raw milk was 12.63%. The milk adulterated with 0.1% of detergent had 13.25%, where as with 0.3, 0.5, 0.7 and

TABLE 17.1

Change in Quality Characteristics of Milk Adulterated with Detergent

Parameters	*Raw Milk*	*Adulterated Milk with Detergent*				
		0.1%	*0.3%*	*0.5%*	*0.7%*	*0.9%*
Moisture	88.35	86.74	86.10	85.03	84.86	83.61
Fat	3.1	3.2	3.2	3.2	3.2	3.2
Acidity	0.1692	0.1125	0.1005	0.0600	0.0285	0.0180
SNF	8.53	10.05	10.68	11.66	11.83	13.68
Total solids	11.63	13.25	13.88	14.96	15.13	16.98
pH	6.65	6.68	7.06	7.36	7.76	7.84
Specific gravity	1.030	1.030	1.0305	1.0315	1.0320	1.0325
R.I.	1.344	1.345	1.346	1.347	1.348	1.350
Casein	2.62	2.89	3.03	3.35	3.35	3.58
Ash	0.727	0.736	0.918	1.087	1.209	1.231
Lactose	4.7	4.62	4.61	4.51	4.39	4.32

0.9% of detergent, the total solid content were increased like 13.88, 14.96, 15.13, and 16.98%, respectively. R.I. reading increased as 1.345, 1.346, 1.347, 1.348 and 1.350 with 0.1, 0.3, 0.5, 0.7 and 0.9% adulteration respectively. As the concentration of detergent increased the thickness of the milk was also increased. The pH of the raw milk was found as 6.65, upon addition with detergent, the pH fluctuations were recorded as 6.68 7.06, 7.36, 7.76, and 7.84% respectively for adulteration levels.

Adulterated Milk with Neutralizer

Neutralizers such as alkali carbonate or even hydroxides are added by lane man during storage of synthetic milk to reload the developing acidity. Sodium bicarbonate causes digestive disorders and known to destroy ascorbic acid (vitamin C) (Amit kumar et al, 2008). Milk was adulterated with 0.1%, 0.5%, 1% and 2% of neutralizer. The addition has decreased the moisture content and increased the total solids. pH is also increased, whereas percent ash decreased compared to the raw milk (0.727%).

TABLE 17.2

Change in Quality Characteristics of Milk Adulterated with Neutralizer

Parameters	*Raw Milk*	*Adulterated Milk with Neutralizer*			
		0.1%	*0.5%*	*1%*	*2%*
Moisture	88.35	88.22	87.81	87.75	87.23
Fat	3.1	3.1	3.1	3.2	3.2
Acidity	0.1692	0.142	0.123	0.128	0.126
SNF	8.53	8.53	8.76	9.51	9.51
Total solids	11.63	11.63	11.86	12.71	12.71
pH	6.65	6.8	6.89	6.89	6.92
Specific gravity	1.030	1.030	1.032	1.035	1.035
R.I	1.344	1.344	1.345	1.345	1.345
Casein	2.62	2.62	2.76	2.89	2.89
Ash	0.727	0.623	0.638	0.658	0.663
Lactose	4.7	4.73	4.78	4.81	4.84

Adulterated Milk with Urea

Urea is an indirect adulterant in milk. It usually comes from synthetic milk. In the study, milk was adulterated with 0.5, 1, 3, 5 and 7% of urea. Table 17.3 revealed that the addition has increased the total solids content and reduced the moisture from 88.35% to 82.48% for 7% adulterated sample. Variation was also observed in the casein percent in milk. The normal casein content of raw milk was 2.62%, and casein content of milk adulterated with 0.5,1, 3, 5 and 7% of urea was found increasing like 2.805, 2.89, 3.06, 3.23 and 3.623% respectively.

TABLE 17.3

Change in Quality Characteristics of Milk Adulterated with Urea

Parameters	*Raw Milk*	*Adulterated Milk with Urea*				
		0.5%	*1%*	*3%*	*5%*	*7%*
Moisture	88.35	87.66	87.55	86.26	84.77	82.48
Fat	3.1	3.2	3.2	3.2	3.2	3.2
Acidity	0.1692	0.1530	0.1485	0.1422	0.1395	0.1125
SNF	8.53	8.79	9.35	10.85	12.37	13.49
Total solids	11.63	11.99	12.55	14.05	15.57	16.69
pH	6.65	6.68	6.79	7.21	7.64	8.35
Specific gravity	1.030	1.0313	1.0325	1.035	1.035	1.035
R.I	1.344	1.355	1.358	1.365	1.368	1.460
Casein	2.62	2.85	2.89	3.06	3.23	3.23
Ash	0.727	0.694	0.668	0.667	0.659	0.654
Lactose	4.7	4.74	4.85	5.06	5.15	5.21

Color

L*, a*, b* values for the milk adulterated with detergent at different concentration were measured using hunter and were tabulated in Table 17.4.

As per Table 17.4, Hunter color coordinate pattern clearly revealed a change in the milk color after adulteration. *L* values showed a gradual decrease after the addition of adulterants indicating a drift in the color coordinate values from lightness to darkness (*L* = 100 - pure white, *L* = 0 - pure black). Lightness of control milk

TABLE 17.4

Change in Colour Values of Adulterated Milk

Samples		*L**	*a**	*b**
Control		92.70	−1.31	13.49
Milk adulterated	0.1	91.82	−1.37	12.98
with detergent (%)	0.3	91.22	−1.24	12.76
	0.5	90.56	−1.42	12.54
	0.7	90.16	−1.74	12.33
	0.9	89.72	−1.81	12.11
Adulterated milk	0.1	90.85	−1.83	13.26
with neutralizer (%)	0.5	90.69	−1.74	13.13
	1.0	90.10	−1.71	13.09
	2.0	89.75	−1.70	12.60
Adulterated milk	0.5	92.18	−1.31	13.55
with urea (%)	1.0	92.06	−1.26	13.63
	2.0	91.76	−1.33	13.79
	3.0	91.37	−1.37	14.10
	5.0	90.68	−1.47	14.19

was 92.70 whereas adulteration with 0.9 % detergent and 2.0 % neutralizer brought a decrease in the lightness to 89.72 and 89.75 respectively. Lightness of milk was more affected in samples adulterated with neutralizer and low quantities of added neutralizer were sufficient enough to bring to down the natural color of milk when compared to the effect of other adulterants at the same quantities. The a* values of adulterated samples showed different reactions to the adulteration substances and concentration. Values showed decrease in the samples adulterated with detergent and urea whereas showed an increase with the increasing concentrations of neutralizer. The *b** values representing the yellowness of milk decreased with the addition of detergent and neutralizer whereas showed an increase in the samples adulterated with urea.

Microbial Analysis

Adulterated Milk with Detergent

The microbial load in raw milk was found to be 2.6 × 10^3 cfu/g. Milk adulterated with 0.1% detergent was 37 × 10^2cfu/g, where as the milk adulterated with 0.3, 0.5, 0.7 and 9% showed a microbial load of 19 × $10,^2$ 5 × 10^1, 2 × 10^1 and 8 × 10^1 cfu/g respectively. This indicates the growth of colonies were decreased, with increasing percentage of detergent, when compared to raw milk. Yeast and molds was not found in adulterated milk samples. The coliform count in raw milk batch was found as 1.2 × 10^1cfu/g and adulterated milk with detergent was found to be decreased from 4 × 10^1 at 0.1% to 2 × 10^1 with 0.3% adulterated. The coliform count also nil in all adulterated milk samples.

Milk Adulterated with Neutralizers

Milk adulterated with 0.1% neutralizer was 6.0 × 10^3 Colonies where as the milk adulterated with 0.5, 1 and 2% of neutralizer showed 4.5 × 10^3, 5 × 10^1, and 3 × 10^1 cfu/g respectively. This indicates the growth of colonies was decreased, with increasing percentage of neutralizers, when compared to raw milk. The coliform of adulterated milk with neutralizer batch was found as nil at 0.1 and 0.5%. Yeast and molds also decreased with increasing levels of neutralizers, from 1 × 10^1 to nil at 0.1 and 2% respectively.

Milk Adulterated with Urea

Milk samples adulterated with 0.5% urea was 8.2 × 10^2 cfu/g. whereas the milk adulterated with 1, 3, 5 and 7% of urea showed 3.05 × 10^2, 8 × $10,^1$ 7 × 10^1 and 4 × 10^1 cfu/g respectively. This indicates the growths of colonies were decreased, with increasing percentage of urea, when compared to raw milk. The coliform count in the milk batch was found as nil while yeast and molds counts were 3 × 10^1 cfu/g. The coliform count of the adulterated milk with urea was found to be decreasing from 5.5 × 10^1 cfu/g at 0.5% to 3.0 × 10^1 cfu/g with 7% adulteration.

Yeast and molds also decreased with increasing in urea adulteration from 2.5×10^1 cfu/g to nil at 0.5 and 7% respectively.

Conclusion

Study revealed a significant and direct relation between various physico-chemical and microbiological quality characteristics of milk against different adulterants and their concentration levels. Urea, neutralizers and detergents when added in fresh milk affected its natural and over all composition and in turn contaminated the sample with hazardous constituents, as the prolonged accumulation of the chemicals in the form of synthetic milk causes various disorder and disease in human beings.

Based on the above studies, milk testing strips were developed to detect the presence of added adulterants in milk. The strips are cheap, convenient, dependable and accurate.

References

Amit Kumar, Vivek Sharma, Dharshan Lal and V. Unnikrishana, Indian Dairy 2008, 43-46.

Omkar kumar and T.Rai, *Indian Food Industry* Volume -29, 2010.

A BI (1979), *Method for Determination of Fat by the Gerber Method*, IS: 1224, Bureau of Indian Stards, New-Delhi.

Aneja R.P., Mathur B.N., Chan R.C. Banerjee A.K. (2002). *Technology of Indian Milk Products*. Traditional Milk Products of India. Indian Dairyman, 43(9) 407-413.

Aneja V.P., Rajorhia G.S Mathur S.K. (1982). Asian Journal Dairy Research, 1, 41-44.

Anwar Sadat, Pervez Mustajab Iqbal A. Khan (2005) *Determining the Adulteration of Natural Milk with Synthetic Milk using Acconductance Measurement.*

AOAC, (1973) *Sampling Analysis of Commercial Fats Oils* 8 (53) .

AOAC, (1990, 1997) *Official Method of Analysis*, Association of Analytical Chemists, Washington D.C.

Baisya R.K., (2006). *Challenges of Marketing Traditional Indian Heritage Foods*, Indian Food Industry, 25(6) 69-70.

Bazinet L., Ippersiel D., Gendron C, Mahdavi B, Amiot J, Lamarche F. (2006) *Effect of Added Salt increase in Ionic Strength on Skim Milk Electro Acidification Performances.* Journal of food science technology. 26(5) 259-264.

Bhaskaracharya, R.K.; Shah, N.P. (2001) *Texture of Mozzarella Cheeses*, Australian Journal of Dairy Technology, 56(1) 9-14.

Borkova, -M; Snaselova, -J (2005) *Possibilities of different Animal Milk Detection in Milk Dairy Products*. Czech-Journal-of-Food-Sciences, 23(2): 41-50.

Brusilovskii, L.P.; Shidlovskaya, V.P. (1996) *Ionometric Detection of Milk Adulteration with Soda Ammonium Compounds*. Journal-Article Molochnaya-. (1): 19-21.

BS ISO 17792:2006, United Kingdom, British Stards Institution (2008) Milk, Milk Products Mesophilic Starter Cultures. Enumeraton of Citrate-Fermenting Lactic Acid Bacteria. Colony-count Technique at 25°C. British Stard. 22pp.

Chin,-X-W-D; Jacobs,-C; Hammer,-P (2007) *Epidemiology of Staphylococcus Aureus in a Spray Drying Dairy Plant.* Journal Article, Kieler-Milchwirtschaftliche-Forschungsberichte, 59(3): 181-189.

Cordeiro,-F; Bordin,-G; Rodriguez,-A-R; Hart,-J-P (2001) *Uncertainty Estimation on the Quantification of Major Milk Proteins by Liquid Chromatography.* Journal-Article Analyst `126(11): 2178-2185, Department of Electronics Engineering, Aligarh Muslim University, Aligarh, U.P. India.

DD ISO/TS 6733:2006, United Kingdom, British Stards Institution (2008)

Determination of Lead Content. Graphite Furnace Atomic Absorption Spectrometric Method. British Stard. 2008, 22pp.

Di Wu, Yong He, Shuijuan Feng, Da-Wen Sun (2007), *Study on Infrared Spectroscopy Technique for Fast Measurement of Protein Content in Milk Powder* based on LS-SVM, Journal of Food Science, 39(4) 484-485.

FDA Centre for Food Safety Applied Nutrition (2003). Grade A Pasteurized Milk ordinance.

Grant, I.R.; Rowe, M.T. (2001) *Methods for Detection Enumeration of viable Mycobacterium Paratuberculosis from Milk Milk Products*, Bulletin of the International Dairy Federation, No: 362, 41-52.

Grigioni,-G; Biolatto,-A; Irurueta,-M; Sancho,-A-M; Paez,-R; Pensel,-N (2007) *Color Changes of Milk Powder due to Heat Treatments Season of Manufacture.* Ciencia-y-Tecnologia-Alimentaria. 5(5): 335-339.

Harding,-F (1990) *Milk Adulteration-Freezing Point Depression.* Journal-of-the-Society-of-Dairy-Technology. 1990; 43(3): 61.

Harju, M (2001) *Recent Developments in the Improvements of the Utilization of Lactose Minerals obtained from Milk as Food Ingredients*, International Journal of Dairy Technology, 54(2) 61-63.

Horiuchi,-H; Inoue,-N (2008) *Delicious Fermented Milk with Low Milk Fat Content Process for Producing the same*, Meiji Dairies Corp.

Indian Stard Methods of Test for Dairy Industry, IS: 1479 (part II) -1961.

ISO 8069:2005, Belgium, Bureau de Normalisation (2007) Dried milk. Determination of content of lactic acid lactates. Belgian Stard. NBN EN ISO 8069:2007, p-24.

Izquierdo,-F-J; Penas,-E; Baeza,-M-L; Gomez,-R (2008) *Effects of Combined Microwave Enzymatic Treatments on the Hydrolysis Immunoreactivity of Dairy Whey Proteins.* International-Dairy-Journal. 2008; 18(9): 918-922.

Jean-François Fairise Philippe Cayot (1998) *New Ultrarapid Method for the Separation of Milk Proteins by Capillary Electrophoresis*, Journal of Agriculture Food Chem., 37, 44.

Karadjova,I,; Girousi,S,; Iliadou, E.; Stratis, I. (2000) *Determination of Mineral Content in Milk Milk Products*, Mikrochimica Acta 134 (3/4) 185-191.

Katserikova, N.V.; Korotkaya, E.V.; Pozdnya-Kovskii, V.M. (2000) *â-Carotene for Fortified Dairy Products*, Molochnaya Promyshelnnost, 3, 37-39.

Kulazhanov,-K-S; Vykhrest,-N-Y; Baidinaeva,-A-T (2007) *Non-thermal Milk Treatment Method.* Aktsionernoe Ovschestvo 'Almatinskii Tekhnologicheskii Universitet'. Journal of Food Science Technology, 29(5) 298-300.

L. Szijarto (2003) *Determination of Added Water Bovine Milk to Caprine Milk*, Journal of Dairy science., 44(3) 232-235.

M F Mabrook, A M Darbyshire M C Petty (2006) *Quality Control of Dairy Products using Single Frequency Admittance Measurements*, Meas. Sci. Technol. 17 275-280, University of Durham, UK.

M. F. Mabrook M. C. Petty (2003) *A Novel Technique for the Detection of Added Water to Full Fat Milk using Single Frequency Admittance Measurements*, Journal of Food Science Technology, 54(2) 78-80 University of Durham, UK.

M.B. Grufferty P.F. Fox (1985), *Department of Food Chemistry*, FST Journal, 9:1-9.

M.I. Abu-Taha A.M. Ayyad (2006) *Monitoring Some Liquid Cladding Layer Properties using LED Beam Transmittance Coefficients through the Fiber itself*, Journal of Dairy science, 44(5) 327-332.

Marsili, R.T. (2000) *Shelf Life Prediction of Processed Milk*, Journal of Agricultural Food Chemistry 48 (8), 3470-3475.

Milk Processing Guide Series, Vol-2, published by- FAO/TCP/KEN/6611 Project, Small Scale Dairy Sector Training Institute-Naivasha.

Milk Quality, F. Harding Frank (1995), Publishers Chapman Hall, Newyork.

Munir-Khan; Rajah,-K-K; Haines,-M (1999) *Quantitative Techniques in the Measurement of Milk Adulteration in Peshawar*, Pakistan. International-Journal-of-Dairy-Technology 52(1): 20-25.

Ogawa,-A (2007) *Emulsion Stabilizer Milk Beverage*, Mitsubishi Chemicals Corp. Journal of Dairy Food Science, 40(1) A15-A21 Press, New Delhi.

Okazaki, Y.; Utsumi, S. (2000) *Forces Involved in the Formation maintenance of Curd prepared with Soyabean Globulins followed by Fermentation of Milk using Lactic Acid Bacteria*, Journal of the Japanese Society for FST, 48(5) 356-360.

Page Pederson (2002), Volume-2, Issue-2, the lab ledger.

Paradkar,-M-M; Singhal,-R-S; Kulkarni,-P-R (2000) *An Approach to the Detection of Synthetic Milk in Dairy Milk Detection of Detergents*. International-Journal-of-Dairy-Technology. 53(3): 92-9.

Paradkar,-M-M; Singhal,-R-S; Kulkarni,-P-R (2000) *An Approach to the Detection of Synthetic Milk in Dairy Milk Detection of Urea*. International-Journal-of-Dairy-Technology, 53(3): 87-91.

Ruijia Yang, Wei Huang, Lishi Zhang, Miles Thomas and Xiaofang Pei (2006) *Milk Adulteration with Melamine in China*: Crisis Response, West China School of Public Health, Sichuan University, Chengdu, China. Publication.

S. Barbut (1996). *Use of Fibre Optics to Study the Transition from Clear to Opaque whey Protein Gels*, International Journal of Dairy Science, 8, 173-176.

OO

18

Experimental Investigation of Convective Drying: Measuring Equilibrium Moisture Content and Drying Time of Moist Potato

V.P. Chandra Mohan and Prabal Talukdar

Abstract

The micro-organisms and bacteria lead the decomposition/decay of moist potato. These failures are almost negligible when the moisture content of potato is reduced below 10%. Convective drying effectively prevents those defects/failures in the food stuffs. An experimental facility is recently developed at the Indian Institute of Technology Delhi for convective drying of moist object. Transient moisture content of a rectangular shaped moist potato ($4 \times 2 \times 2$ cm^3) is measured by this test facility. Initial moisture content is measured by a thermostatically controlled hot air oven. Moisture content is measured at different air temperatures of 40, 50, 60 and 70°C with air velocities of 2, 4 and 6 m/s. The equilibrium moisture content and equilibrium drying time (Fourier number) is found and tabulated. A correlation is developed for equilibrium moisture content and Fourier number based on the drying air velocity and temperature. Density of potato is estimated from the experimental data and a good agreement is noticed when compared with the experimental data from literature.

Introduction

Convective drying is one of the oldest methods for preserving food products. It inhibits microbial development and quality decay in food products. It is a simultaneous heat and mass transfer processes and removes moisture content from the object. In convective drying, it is necessary to know the equilibrium moisture content and equilibrium drying time. These two parameters are mainly used to determine the desirable conditions of velocity and air drying temperature, to predict the energy consumed during drying, and also used to predict the quantity of manual work required for convective drying. Finding the equilibrium moisture content and equilibrium drying time of moist potato is a big task during experiments. Because these two parameters depends on air drying temperature, air flow velocity, humidity of supplied air, size of the moist object and quantity of moisture content to be removed. Also, experiments in convective drying can enhance the understanding of the drying parameters like density of moist potato and drying rate. A well-designed and well-controlled experimental facility is necessary to carry out such an analysis.

In recent years, a large number literature is seen on convective drying. Velic et al.[1] investigated experimentally the convective drying of apple in laboratory conditions and investigated the influence of airflow velocities on drying kinetics and heat transfer coefficient. The structural changes of potato by light microscopy and effect of shrinkage during drying were analyzed experimentally by Wang and Brennan[2]. The effect of shrinkage during convective drying was analyzed by Simal et al.[3]. Hussain and Dincer[4] numerically predicted the temperature and moisture content distribution of a two dimensional cylindrical moist object. Dutta et al.[5] investigated experimentally the drying behavior of a spherical object and validated their 1-D numerical model with experimental data. Chandramohan and Talukdar[6] have developed a numerical model for estimating the temperature and moisture distribution in a 3-D moist object with variable heat and mass transfer coefficients. The coefficients are calculated from a Computational Fluid Dynamics (CFD) model.

Variety of experimental works in convective drying of moist potato found in the literatures. But, there is no correlation found about equilibrium moisture content and equilibrium drying time with atmospheric drying conditions, which is very essential to drying industries and researchers. A properly designed experimental set up with known input and output conditions is required for such type of effective experiments. Also, experimentally it is essential to study the drying parameters like air flow velocity, air drying temperature, amount of moisture content of the object, density and drying rate. This motivates the present experimental work of designing and developing a well designed experimental setup and drying experiments from same.

This study consists of following essential parts, (i) finding the initial moisture content of the moist object, (ii) calculating time dependent moisture content of moist object, (iii) finding the equilibrium moisture content and mass transfer Fourier number at different drying air velocity and temperature conditions, (iv) making correlations by non linear regression analysis, (v) calculating transient density of potato with various air flow velocity and air drying temperatures and (vi) validating the experimental results with results from literatures.

Experimental Setup

Materials and Methods

The skin of a fresh potato is removed and then cut into slabs with a length (L) of 4 cm, breadth (B) of 2 cm and a width (H) of 2 cm. Potato sample is placed on a tray placed in the test section. A similar dimension of potato piece is prepared for finding the initial moisture content by hot air oven.

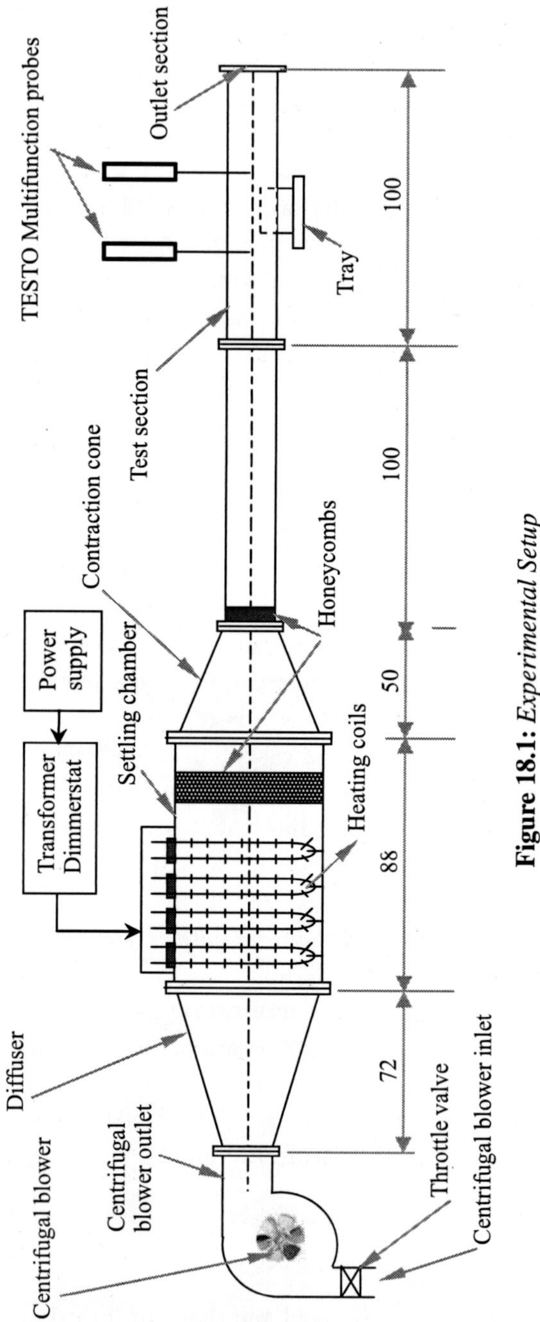

Figure 18.1: *Experimental Setup*

Experimental Design

An experimental setup for convective drying is built at IIT Delhi. This consists of an inlet section, a convergent and divergent section, a settling chamber, a test section and an outlet section. The maximum mass flow rate is fixed at 0.1 m^3/s. Based on the design calculations, the actual pressure drop is calculated as 1346.74 Pa. The power requirement of a centrifugal blower and a motor is calculated as 0.353 HP to work against this pressure drop. Hence, a 0.5 HP capacity centrifugal blower is selected.

Figure 18.1 shows the block diagram of the experimental setup. Honeycombs are used in settling chamber and entrance of test section for flow straightening purpose. U type of heaters (10 numbers, each 500 W) are fixed in the settling chamber in a staggered manner. Auto Transformer Dimmerstat portable type (15D-3P) is connected for maintaining/controlling the voltage supply of heating coils thus maintains the air temperature. The multi function hygrometer probes (TESTO, Part no: 0635 1535) are inserted at the top of test section which is used to measure the upstream and downstream temperature, %RH and velocity of air. Both hygrometer probes are connected to TESTO indicators, Part no: 435-1.

Experimental Procedure

The air velocities chosen for the present study are 2, 4 and 6 m/s. The temperature is maintained at the range of 40 to 70°C in the test section. The centrifugal blower and heaters are first switched ON for 20 minutes. After achieving the required velocity and temperature conditions, a known initial mass of rectangular shaped moist potato (4 cm × 2 cm × 2 cm) is placed on the thin tray at the middle of the test section. Mass of the potato, upstream and downstream temperature and humidity is measured for every 20 minutes of 16 hours of drying. Mass of the potato is measured by OHAUS's model Adventurer Basic level Electronic Single Pan Balance (Model AR 3130) with a readability of 0.001 g.

For finding the initial moisture content, a thermostatically controlled hot air oven is used. A rectangular shaped moist potato (4 cm × 2 cm × 2 cm) is covered with an aluminum paper and is kept in a hot air oven (YSI – 431, IS – 3119) where the temperature is maintained at 105°C. Mass of the moist object is measured at every 3 hours. It is noted that initial moisture content value is almost same from both procedure. It is noted that 83% of initial moisture $\left(\text{Initial moisture content} = \frac{m_{initial} - m_{final}}{m_{initial}} \times 100\%\right)$ in terms of mass is existing in the moist potato.

Results and Discussion

The experimental density is calculated from volume and mass of the moist object. The initial mass of moist potato (m_{wet}) and final mass of the dry potato

(m_{dry}) are measured with the weighing balance. The moisture content in dry basis is calculated from the expression,

$$M_0 = \frac{m_{wet} - m_{dry}}{m_{dry}} \quad (1)$$

The initial moisture content is found to be $M_0 = 4.8862$ kg/kg of dry basis (db). In a similar way using Eq.(1), the transient moisture content of the moist object is calculated at different times. Experimental density is calculated from the mass of the moist object divided by the volume of the moist object. Figure 18.2 shows the plot between density and moisture content of the moist object at 40ºC. Experimental data are compared with Wang and Brennan's [2] correlation given as:

$$\rho = A_1 + A_2 \exp (A_3 M^2) \quad (2)$$

where, A_1, A_2 and A_3 are constants and found from non linear regression procedure and M is the moisture content. Good agreement between present experimental data and Wang and Brennan's correlation data can be seen from the figure.

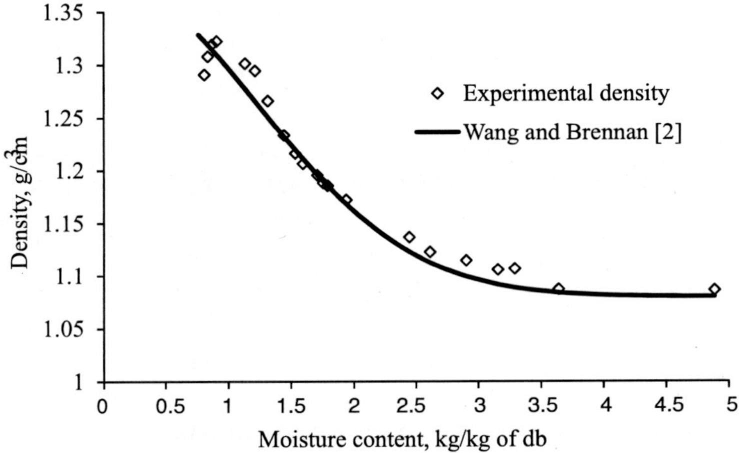

Figure 18.2. *Comparison of Experimental Results with Results from Literature at 2m/s: Variation of Density with Moisture Content of the Moist Object at 40°C*

During the experiments at a particular air velocity and temperature, the product loses its moisture content gradually when the drying time is increased. After a particular drying time there is no moisture transfer to air as the moisture content in the object and dry air is almost equal. This moisture content is called equilibrium moisture content and the corresponding drying time is called equilibrium drying time. The equilibrium drying time is represented by a non dimensional number called equilibrium Fourier number ($\tau_e = D_e t_e / L_d^2$), where, D_e = equilibrium diffusion coefficient, t_e = equilibrium drying time and L_d = hydraulic diameter. The equilibrium moisture content and equilibrium Fourier number at different air velocities and air drying temperatures are tabulated in Table 18.1.

TABLE 18.1

Equilibrium Moisture Content and Fourier Number

Vel, m/s	*Temp, °C*	*Equilibrium Moisture content, M_{eq} (kg/kg of db)*	*Equilibrium Fourier number (τ_e)*
2	40	0.761	0.198344
	50	0.381	0.268847
	60	0.236	0.274528
	70	0.1987	0.302432
4	40	0.2496	0.31981
	50	0.1752	0.335876
	60	0.093	0.394828
	70	0.043	0.413882
6	40	0.12	0.360966
	50	0.089	0.390767
	60	0.055	0.411812
	70	0.041	0.422116

The equilibrium moisture content of potato is decreased as the air drying temperature and air velocity is increased. The equilibrium Fourier number is calculated based on the hydraulic diameter of potato and it helps to estimate the time required to dry a substance under specified conditions. A non linear regression analysis is used to find the correlation of equilibrium moisture content (M_e) and a regression analysis of four parameter polynomial equation is used for finding the correlation for equilibrium Fourier number (τ_e) as given below

$$M_{eq} = 43140.87\, v^{-1.52706}\, T^{-2.685} \quad \ldots\ldots\ldots\ldots (r^2 = 0.99) \tag{3}$$

$$\tau_e = a + (b/v) + (c/v^2) + (d/T), \quad \ldots\ldots\ldots\ldots (r^2 = 0.972) \tag{4}$$

where, a, b, c and d are constants given as: a = 0.6037185, b = – 0.2932325, c = –0.16935 and d = –8.095967, v is velocity in m/s and T is drying air temperature in °C.

The non dimensional moisture content (Φ) is calculated as,

$$\Phi = \frac{M - M_{eq}}{M_0 - M_{eq}} \tag{5}$$

where, M_0 is the initial moisture content and M_{eq} is the equilibrium moisture content of potato.

Figure 18.3(a) shows the effect of non dimensional moisture content of the object with drying time for different drying air temperatures and at air velocity 6 m/s. The product is losing its moisture continuously during the entire drying time.

Also it is noticed that, when the air drying temperature is increased, the drying time of potato is decreased. The temperature inside the object increases with an increase in the drying air temperature as a result of convection heat transfer from the drying medium to the product and heat diffusion inside the product. Also, due to the difference in temperatures between the supplied air and the surface of the moist object, heat is transferred from the supplied air to the object by convection. Thus the moist potato loses more moisture in higher drying temperature condition and hence, the drying time is decreased.

(a)

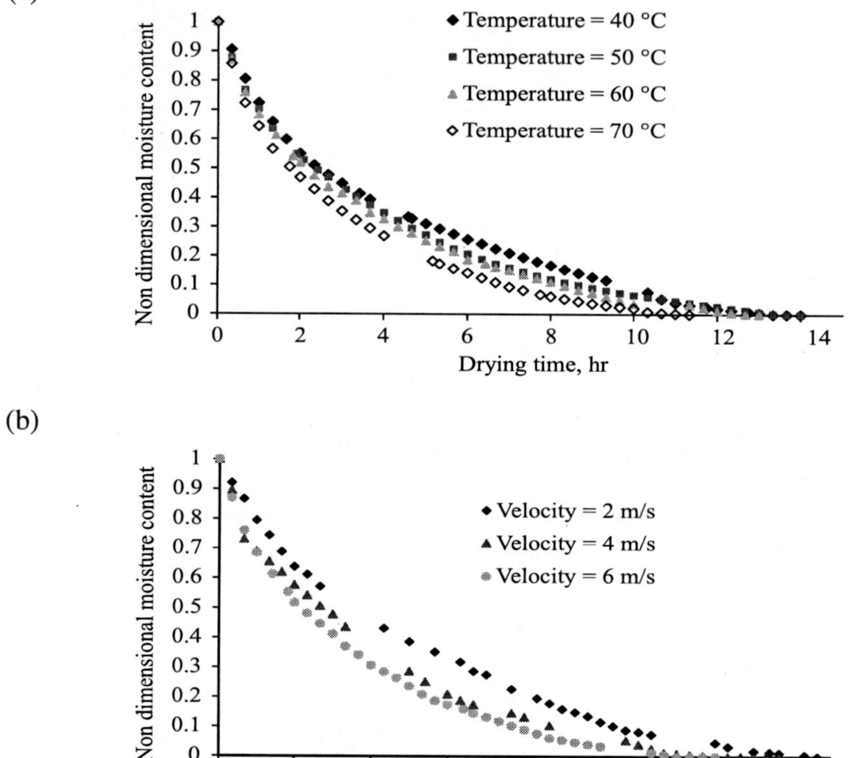

Figure 18.3: *Variation of Non-Dimensional Moisture Content of the Object with Drying Time at (a) Different Temperatures and at Air Velocity of 6m/s and (b) Different Air Velocities and at an Air Temperature of 60°C*

Variation of non dimensional moisture content at different drying air velocities and drying air temperature at 60°C are shown in Figure 18.3(b). The drying rate is higher for air velocity of 4 and 6 m/s as compared to 2 m/s, but there is not much variation found in drying rate between the velocities of 4 and 6 m/s. This is because as the air velocity is increased, the surface transfer coefficient and diffusion coefficient for the drying product are also increased almost proportionally resulting an increase in the rate of moisture transfer. Velic et al. [1] conducted some experimental work for drying of apple and found similar results.

Figure 18.4(a) shows the variation of density with non dimensional moisture content for different air temperatures and at air velocity 6 m/s. It is observed that in the early stage of drying (non dimensional moisture content from 0.6 to 1) the density is almost constant with respect to moisture content. The density starts increasing gradually in the later stage with decrease in moisture content and reaches a maximum value for all the temperature considered in this study. The corresponding non dimensional moisture content at this maximum density is approximately 0.1 as seen from Figure 18.4(a).

(a)

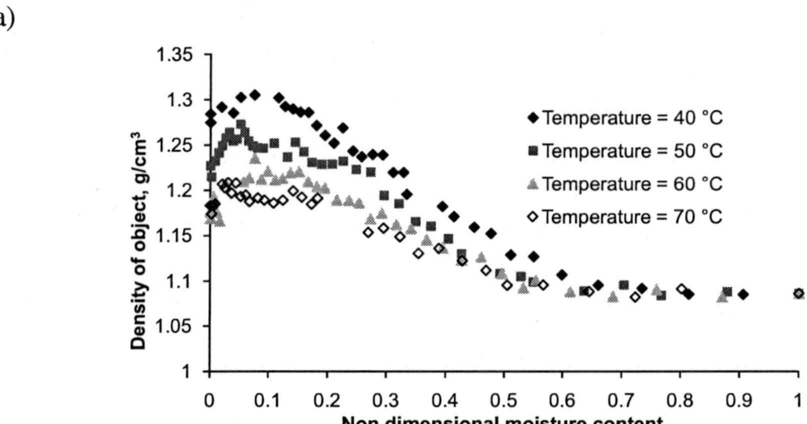

Density variation with respect to non dimensional moisture content at 2 and 4 m/s also exhibits similar trend of variation. Shrinkage rate is responsible for the increased apparent density at non dimensional moisture contents from 0.1 to 0.6 (Figure 18.4a)..

(b)

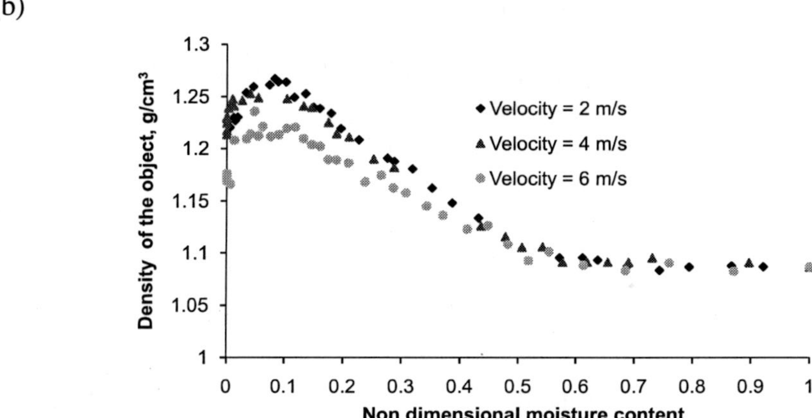

Figure 18.4: *Variation of Density with Non-Dimensional Moisture Content at (a) Different Temperatures and at 6 m/s and (b) Different Velocities and at 60°C*

Further reduction in moisture content resulted in a decrease in the density of the moist object (non dimensional moisture content less than 0.1) because the volume of the sample does not change with the further loss of water. Therefore, the density of the sample decreases with decreasing moisture content at low moisture contents

Figure 18.4(b) shows the density variation with respect to non dimensional moisture content for various velocity conditions at 60°C. The plot shows that at lower moisture content, density is lower for higher velocities. During the initial drying period or the region with high moisture content, a constant density is observed for almost all velocity conditions. For lower moisture content region, at a particular moisture content, the density is decreased with increasing velocities. This is because of the fact that higher air flow velocity eliminates more amount of moisture from the object and thereby, the object loses its mass and density.

Conclusions

An experimental setup is carefully made for convective drying of moist object at Indian Institute of Technology, Delhi. This facility is designed to achieve a maximum velocity of 8 m/s and maximum temperature of 70°C in the test section.

In the experiments performed, the air velocity was kept at 2, 4 and 6 m/s and the temperatures were maintained from 40 to 70°C in the test section. A rectangular shaped potato slice of cross section 4 cm × 2 cm × 2 cm was taken as a sample moist object. The following important conclusions are drawn:

- The initial moisture content of potato was observed to be 83% in terms of mass and it is quantified as 4.8862 kg/kg of db.
- Equilibrium moisture content and equilibrium Fourier numbers are estimated from the experiments and are tabulated.
- Drying correlations relating equilibrium moisture content and Fourier number with air drying velocity and air temperature are achieved by non linear regression analysis.
- The density of potato is almost constant at higher moisture content and is increased further with decrease in moisture content. Density is seen to be decreased with the increase of drying air temperature at a given moisture content.
- The higher drying temperature and air velocity enhances the drying rate of moist potato.

NOMENCLATURE

a, b, c, d	Constants
A_1, A_2, A_3	Constants
B	Breadth of potato, cm
D	Diffusion coefficient, m^2/s

H	Width of potato, cm
db	dry basis
L	Length of the moist object, cm
L_d	Hydraulic diameter, m
m	Mass of the object, g
M	Moisture content, kg/kg of db
t	time, s
T	Temperature, °C
v	velocity, m/s
Greek Symbols:	
Φ	Non dimensional moisture content
ρ	Density of object, g/cm^3
τ	Mass transfer Fourier number
Subscripts:	
eq	equilibrium
0	initial condition

References

D. Velic, M. Planinic, S. Tomas and M. Bili, *Influence of Airflow Velocity on Kinetics of Convection Apple Drying*, Journal of Food Engineering 64 (2004) 97-102.

N. Wang and J.G. Brennan, *Changes in Structure, Density and Porosity of Potato during Dehydration*, Journal of Food Engineering 24(1993) 61-76.

S. Simal, C. Rossello, A. Berna and A. Mulet, *Drying of Shrinking Cylinder-shaped Bodies*, Journal of Food Engineering 37(1998) 423-435.

M.M. Hussain and I. Dincer, *Two-dimensional Heat and Moisture Transfer Analysis of a Cylindrical Moist Object subjected to Drying*: A Finite-difference Approach, International Journal of Heat and Mass Transfer 46(2003), 4033-4039.

S.K. Dutta, V.K. Nema and R.K. Bharadwaj, *Drying Behavior of Spherical Grains.* International Journal of Heat and Mass Transfer 31(4) (1988) 855-861.

V.P.C. Mohan and P. Talukdar, *Three dimensional Numerical Modeling of Simultaneous Heat and Moisture Transfer in a Moist Object subjected to Convective Drying*. International Journal of Heat and Mass Transfer 53(2010) 4638-4650.

OO

Celiac Disease: New Advancements in Detection and Therapy

Shanky Bhat, Swarnabhra Shukla and Preety Pant

Abstract

Celiac disease is recognized as a common multisystemic disorder that may be diagnosed at any age. At this time, the gluten-free diet remains the only available treatment. The gluten-free diet is a complex and challenging diet, but current advances in the food industry are making it easier to follow. Patients who understand the long-term consequences of celiac disease will make informed choices in managing their disease. Dietitians provide the tools that patients need to successfully understand the diet and integrate it into every aspect of their lives, leading to overall improvements in the physical and emotional challenges of the disease. Celiac disease a condition resulting from an inappropriate immune response to the dietary protein gluten. The manifestations of celiac disease range from no symptoms to over malabsorption with involvement of multiple organ systems and an increased risk of some malignancies. When celiac disease is suspected, initial testing for serum immunoglobulin A (IgA) tissue transglutaminase (tTG) antibodies is useful because it offers adequate sensitivity and specificity at a reasonable cost. A positive IgA tTG result should prompt small bowel biopsy with at least four tissue samples to confirm the diagnosis. However, 3 percent of patients with celiac disease have IgA deficiency. Therefore, if the serum IgA tTG result is negative but clinical suspicion for the disease is high, a serum total IgA level may be considered. Screening of asymptomatic patients is not recommended. The basis of treatment for celiac disease is adherence to a gluten-free diet, which may eliminate symptoms within a few months. Patients should also be evaluated for osteoporosis, thyroid dysfunction, and deficiencies in folic acid, vitamin B_{12}, fat-soluble vitamins, and iron, and treated appropriately. Serum IgA tTG levels typically decrease as patients maintain a gluten free diet. With more patients being diagnosed, there is a greater need for health care professionals who are knowledgeable about celiac disease and the glutenfree diet. Expert dietitians are responsible for nutrition assessment, treatment of nutritional deficiencies, and education of patients with celiac disease.

Introduction

Celiac disease is becoming an increasingly recognized autoimmune enteropathy caused by a permanent intolerance to gluten. Once thought to be a rare disease of childhood characterized by diarrhea, celiac disease is actually a multisystemic disorder that occurs as a result of an immune response to ingested gluten in genetically predisposed individuals. It is a genetically determined autoimmune-like disorder induced by Gluten, the storage protein of wheat and similar proteins found in barley and rye. The presence of gluten in these subjects leads to self-perpetuating mucosal damage, whereas elimination of gluten results in full mucosal recovery. The symptoms of celiac disease vary among individuals and depend on the amount of gluten a person consumes. Symptoms can affect the digestive tract as well as other parts of the body. Common symptoms include excessive gas, abdominal bloating, diarrhea, weight loss, and fatigue. Some people who have celiac disease may have no symptoms. Celiac disease, also known as celiac sprue or gluten-sensitive enteropathy, is far more common than once believed. It commonly runs in families and in populations with other autoimmune diseases and genetic disorders (Source: NIDDK).

Recognizing celiac disease can be difficult because some of its symptoms are similar to those of other diseases. Celiac disease can be confused with irritable bowel syndrome, iron-deficiency anemia caused by menstrual blood loss, inflammatory bowel disease, diverticulitis, intestinal infections, and chronic fatigue syndrome. As a result, celiac disease has long been under diagnosed or misdiagnosed. Due to lack of sensitivity and specificity of antibodies to gliadin component of gluten, it is not now used for diagnostic purposes[2]. Whereas, the endomysial antibody (EMA) has a good specificity for celiac disease and has become the "gold standard "serological test[3]. There was the study which provides substantial evidence for a very strong genetic component in coeliac disease, which is only partially due to the HLA region[1-3].

Epidemiology

In monozygotic twins the likelihood of having celiac disease increases to 75% and 10% to 20% in persons who have a first-degree relative with the disease[4,5]. Further, patients who have Turner's syndrome, Down syndrome, type 1 diabetes mellitus, or other disorders are at increased risk for developing celiac disease[4-8]. In recent years population studies have revealed a much higher prevalence, particularly in individuals of European ancestry.[4, 6, 10] Celiac disease also occurs in people who are not of European descent, although the prevalence is not as great. People from India like those of regions Punjab and Gujarat who lived in England, they developed celiac disease 2.7 times as often as Europeans on a gluten-rich diet; further during the summer months, when wheat commonly replaces maize in the diet a disorder termed *summer diarrhea* has long plagued people of the tropics[6,9].

The prevalence of celiac disease is rising. As a result there is increasing interest in the associated mortality and morbidity of the disease. Screening of asymptomatic individuals in the general population is not currently recommended; instead, a strategy of case finding is the preferred approach, taking into account the myriad modes of presentation of celiac disease. Although a gluten-free diet is the treatment of choice in symptomatic patients with celiac disease, there is no consensus on whether institution of a gluten-free diet will improve the quality of life in asymptomatic screen-detected celiac disease patients. A review of the studies that have been performed on this subject is presented. Certain patient groups such as those with autoimmune diseases may be offered screening in the context of an informed discussion regarding the potential benefits, with the caveat that the data on this issue are sparse. Active case finding seems to be the most prudent option in most clinical situations[25].

TABLE 19.1

Risk Factors for Celiac Disease

Risk Factor	*Prevalence of Celiac Disease among those with Risk Factor (%)*
First-degree Relative with Celiac Disease	5 to 22
Dermatitis Herpetiformis	100
Autoimmune thyroid disease	1.5 to 14
Turner's syndrome	2 to 10
Down syndrome	5 to 12
Type 1 diabetes mellitus	
Children	3 to 8
Adults	2 to 5

Reference 11-15

How to Diagnose CD?

The diagnosis of CD is based on 3 key parameters:

(1) Case identification,

(2) Screening tests, and

(3) Definitive tests.

Diagnosis of Celiac Disease

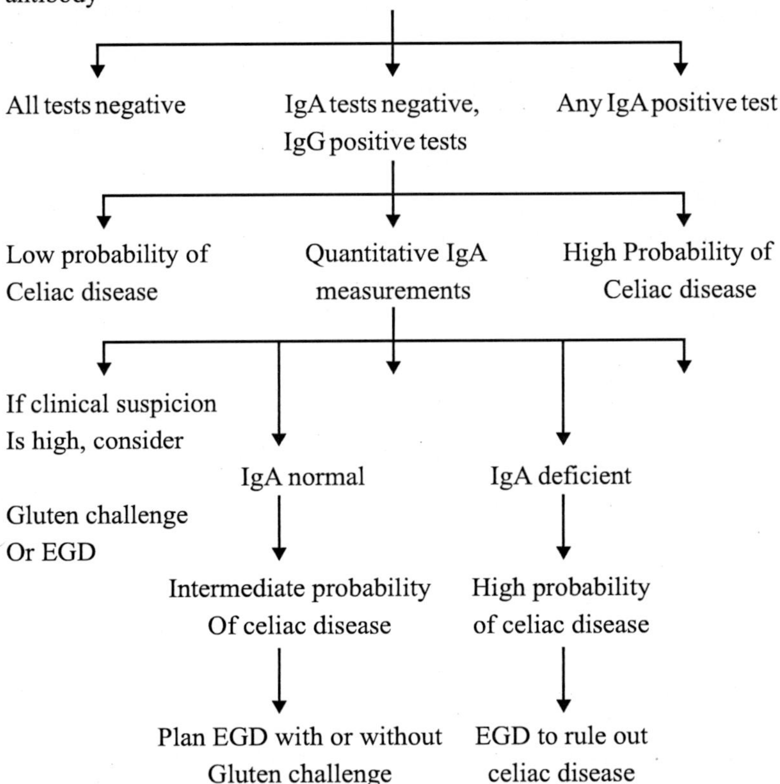

Figure 19.1: *Algorithm for the Diagnosis of Celiac Disease. (EGD: esophagogastroduodenoscopy).*

Development of Serologic Tests

Serum immunoglobulin A (IgA) endomysial antibodies and IgA tissue transglutaminase (tTG) antibodies are the most common serologic markers used for celiac disease screening [17–19]. Because of the low sensitivity and specificity for celiac disease testing for gliadin antibodies is no longer recommended. It has been found that the sensitivity and specificity of testing for IgA endomysial and tTG antibodies are greater than 95 percent[20,21]. But, the sensitivity depends on the degree of mucosal involvement. Since tTG is an autoantigen which is recognized by the endomysial antibody, there is rarely a need to perform both tests[20]. The tTG antibody test uses an enzyme-linked immunosorbent assay, hence it is less costly; in the primary care setting, it is the recommended single serologic test for celiac disease screening[20-22]. The enzyme tissue transglutaminase 2 (tTG) is the main autoantigen for the EMA 5 (endomysial antibody) the recognition of this enzyme allowed development of automated enzyme linked immunoassays (ELISA). Guinea pig tTG was used as an antigen initially; subsequently either recombinant or derived human tTG from red blood cells (H-tTG), was used in the assays. The IgA tTG antibody test has greater than 90% sensitivity and specificity for celiac disease shown by the systematic review of the available studies[2]. A false-negative serologic test result may be there in approximately 3 percent of patients with celiac disease who have IgA deficiency[15]. Celiac disease is still suspected if IgA deficiency is suspected or if serum tTG is negative then the American Gastroenterological Association recommends measuring total IgA levels[12]. Because serologic markers may have false-positive or false-negative results, they cannot be relied on for the diagnosis of celiac disease. However, positive serologic markers can indicate the need for further evaluation with small bowel biopsy, particularly in patients at increased risk. Conversely, negative serologic markers in low-risk patients without IgA deficiency have a high negative predictive value, and small bowel biopsy generally is not needed. Negative markers should never prevent small bowel biopsy if the index of suspicion for celiac disease is high.

Future Therapeutic Options

The quality of life for patients with celiac disease could potentially be improved if a treatment was available that would allow patients to consume gluten, even in small amounts or over a short period of time. Effective treatments for celiac disease, other than dietary restrictions, are being developed[14].

Management

Once the diagnosis of celiac disease has been made, patients should be evaluated for known manifestations and complications *(Table 2)*. Iron deficiency should be treated with supplemental iron. Osteoporosis should be treated with calcium and vitamin D replacement. Depending on individual factors, patients with gluten-sensitive enteropathy may need to take a multivitamin, iron, calcium, magnesium, zinc, selenium, vitamin D, or other nutrients.

TABLE 19.2

Laboratory Evaluations of Patients with Newly Diagnosed Celiac Disease

Hematology
Complete blood cell count
Platelet count
Laboratory tests
Iron level, total iron-binding capacity determination, ferritin level*
Vitamin B_{12} and folate levels
Calcium and phosphate levels
Alkaline phosphatase level
Blood urea nitrogen and creatinine levels
Albumin and total serum protein levels
Aspartate transaminase and alanine transaminase levels
Imaging
Dual energy x-ray absorptiometry (DEXA) of spine and hip
Serologic tests†
Quantitative IgA antiendomysial antibody or quantitative IgA antitransglutaminase

*—*The ferritin level may be misleading: it may be elevated as an acute-phase reactant.*

†—*Serologic testing should be recommended for first-degree relatives of patients with newly diagnosed celiac disease*

Diet for Celiac Patients:

Gluten Content of Some Common Foods

Category	*Contains Gluten*	*Usually Gluten Free*
Breads, cereals, rice, and pasta	Bread or pasta made from barley, bran, gluten flour, graham flour, oat bran,* rye, wheat-based semolina, spelt, wheat, or wheat germ Cereals made with wheat, rye, barley, or oats, or with malt extract or malt flavorings	Bread, cereals, or pasta made from arrowroot corn, buckwheat, corn, cornmeal, hominy, millet, potato starch, rice, rice bran, sago, soy, or tapioca Puffed corn Rice (brown or white); rice noodles
Vegetables and beans	Creamed or breaded vegetables; some French fries Canned baked beans	Plain, fresh, frozen, or canned vegetables Soybeans
Fruits	Some commercial fruit pie fillings and dried fruit	All fruits
Dairy	Malted milk; some milk drinks and flavored or frozen yogurt	Milk and milk products that do not contain gluten additives

Category	*Contains Gluten*	*Usually Gluten Free*
Meat, poultry, fish, shellfish, eggs, and nuts	Any prepared with barley, oats, rye, wheat, or gluten stabilizers or fillers, including some cold cuts, frankfurters, sandwich spreads, sausages, and canned meats	Plain meat, poultry, fish, and shellfish Cold cuts, frankfurters, sandwich spreads, and sausages that do not contain gluten fillers Eggs; nuts and peanut butter
Snacks and condiments	Many commercial salad dressings, prepared soups, condiments, and sauces	Butter and margarine Honey; jam and jelly; molasses; sugar Coconut; hard candy; marshmallows; meringue; plain chocolate
Beverages	Flavored instant coffees; herbal teas Hot cocoa mixes; nondairy cream substitutes	Pure instant or ground coffee; tea Carbonated drinks; fruit juices

*— *Pure oats are usually well tolerated but patients should be cautioned that oat products are frequently contaminated with gluten during processing.*

NOTE: *This is not a complete list. Patients should read food labels to confirm that food is gluten free. Patients should also check with their pharmacists because some medications contain gluten.*
Information from reference 24.

The following are lists of various foods that do not have gluten, may have gluten and do not contain gluten.

	No Gluten	*May Contain Gluten*	*Does Contain Gluten*
Prebiotic Plant Fiber oligofructose and Inulin	Onion, garlic, leeks, Jerusalem artichokes, asparagus, chicory root, jicama, dandelion, banana, agave, jams, Prebiotin		Wheat, barley, rye
Milk and Milk Products	Whole, low fat, skim, dry, evaporated or condensed milk; buttermilk; cream; whipping cream; Velveeta cheese food; American cheese; all aged cheese such as Cheddar, Swiss, Edam and Parmesan	Sour cream, commercial chocolate milk and drinks, non-dairy creamers, all other cheese products, yogurt	Malted drinks

	No Gluten	*May Contain Gluten*	*Does Contain Gluten*
Meat or Meat Substitutes	100% meat (no grain additives); seafood; poultry (breaded with pure cornmeal, potato flour or rice flour); peanut butter; eggs; dried beans or peas; pork	Meat patties; canned meat; sausages; cold cuts; bologna; hot dogs; stew; hamburger; chili; commercial omelets, souffles, fondue; soy protein meat substitutes	Croquettes, breaded fish, chicken loaves made with bread or bread crumbs, breaded or floured meats, meatloaf, meatballs, pizza, ravioli, any meat or meat substitute, rye, barley, oats, gluten stabilizers
Breads and Grains	Cream of rice; cornmeal; hominy; rice; wild rice; gluten-free noodles; rice wafers; pure corn tortillas; specially prepared breads made with corn, rice, potato, soybean, tapioca, arrowroot, carob, buckwheat, millet, amaranth and quinoa flour	Packaged rice mixes, cornbread, ready-to-eat cereals containing malt flavoring	Breads, buns, rolls, biscuits, muffins, crackers and cereals containing wheat, wheat germ, oats, barley, rye, bran, graham flour, malt; kasha; bulgur; Melba toast; matzo; bread crumbs; pastry; pizza dough; regular noodles, spaghetti, macaroni and other pasta; rusks; dumplings; zwieback; pretzels; prepared mixes for waffles and pancakes; bread stuffing or filling

	No Gluten	*May Contain Gluten*	*Does Contain Gluten*
Fats and Oils	Butter, margarine, vegetable oil, shortening, lard	Salad dressings, non-dairy creamers, mayonnaise	Gravy and cream sauces thickened with flour
Fruits	Plain, fresh, frozen, canned or dried fruit; all fruit juices	Pie fillings, thickened or prepared fruit, fruit fillings	None
Vegetables	Fresh, frozen or canned vegetables; white and sweet potatoes; yams	Vegetables with sauces, commercially prepared vegetables and salads, canned baked beans, pickles, marinated vegetables, commercially seasoned vegetables	Creamed or breaded vegetables; those prepared with wheat, rye, oats, barley or gluten stabilizers
Snacks and Desserts	Brown and white sugar, rennet, fruit whips, gelatin, jelly, jam, honey, molasses, pure cocoa, fruit ice, carob	Custards, puddings, ice cream, ices, sherbet, pie fillings, candies, chocolate, chewing gum, cocoa, potato chips, popcorn	Cakes, cookies, doughnuts, pastries, dumplings, ice cream cones, pies, prepared cake and cookie mixes, pretzels, bread pudding
Beverages	Tea, carbonated beverages (except root beer), fruit juices, mineral and carbonated waters, wines, instant or ground coffee	Cocoa mixes, root beer, chocolate drinks, nutritional supplements, beverage mixes	Postum™, Ovaltine™, malt-containing drinks, cocomalt, beer, ale
Soups	Those made with allowed ingredients	Commercially prepared soups, broths, soup mixes, boullion cubes	Soups thickened with wheat flour or gluten-containing grains; soup containing barley, pasta or noodles

	No Gluten	*May Contain Gluten*	*Does Contain Gluten*
Thickening Agents	Gelatin, arrowroot starch; corn flour germ or bran; potato flour; potato starch flour; rice bran and flour; rice polish; soy flour; tapioca, sago	Commercially prepared soups, broths, soup mixes, boullion cubes	Wheat starch; all flours containing wheat, oats, rye, malt, barley or graham flour; all-purpose flour; white flour; wheat flour; bran; cracker meal; durham flour; wheat germ
Condiments	Gluten-free soy sauce, distilled white vinegar, olives, pickles, relish, ketchup	Flavoring syrups (for pancakes or ice cream), mayonnaise, horseradish, salad dressings, tomato sauces, meat sauce, mustard, taco sauce, soy sauce, chip dips	
Seasonings	Salt, pepper, herbs, flavored extracts, food coloring, cloves, ginger, nutmeg, cinnamon, bicarbonate of soda, baking powder, cream of tartar, monosodium glutamate	Curry powder, seasoning mixes, meat extracts	Synthetic pepper, brewer's yeast (unless prepared with a sugar molasses base), yeast extract (contains barley)
Prescription Products		All medicines - Check with pharmacist or pharmaceutical company	

Sample Menu

Breakfast	*Lunch*	*Dinner*
Cream of rice - 1/2 cup	Baked chicken - 3 oz	Sirloin steak - 3 oz
Skim milk - 1 cup	Rice - 1/2 cup	Baked potato - 1 medium
Banana - 1 medium	Green beans - 1/2 cup	Peas - 1/2 cup
Orange juice - 1/2 cup	Apple juice - 1/2 cup	Fruit gelatin - 1/2 cup
Sugar - 1 tsp	Ice cream - 1/2	Butter - 1 Tbsp
	* Ice cream should be made without wheat stabilizers.	Tea - 1 cup Sugar - 1 tsp

References

Green, P.H., Jabri, B. Coeliac disease. Lancet 2003;362:383-391.

Rostom, A., Dube, C., Cranney, A., Saloojee, N., Sy R, Garritty, C., Sampson, M., ZhangL, Yazdi, F., Mamaladze, V., Pan I, Macneil, J., Mack, D., Patel, D., Moher D. *The Diagnostic Accuracy of Serologic Tests for Celiac Disease: A Systematic Review.* Gastroenterology 2005;128:S38-46.

Kapuscinska, A., Zalewski, T., Chorzelski, T.P., Sulej, J., Beutner, E.H., Kumar, V., Rossi, T. *Disease Specificity and Dynamics of Changes in IgA Class Anti-endomysial Antibodies in Celiac Disease.* J Pediatr Gastroenterol Nutr 1987;6:529-34.

Nelsen, D.A.. *Gluten-sensitive Enteropathy (celiac disease): More Common Than You Think. Am Fam Physician.*2002;66:2259-2266.

Presutti, R.J., Cangemi, J.R., Cassidy, H.D., Hill, D.A. *Celiac Disease. Am Fam Physician.* 2007;76: 1795-1802.

Hill, I.D. *Clinical Manifestations and Diagnosis of Celiac Disease in Children.* Waltham, MA: *UpToDate*; 2008.

Barton, S.H., Murray, J.A. *Celiac Disease and Autoimmunity in the Gut and Elsewhere. Gastroenterol Clin North Am.* 2008;37:411-428.

Fasano, A. Berti, I. Gerarduzzi, T. et al. *Prevalence of Celiac Disease in at-Risk and not-at-risk Groups in the United States: A Large Multicenter Study. Arch Intern Med.* 2003;163:286-292.

Schuppan, D., Dieterich, W. *Pathogenesis, Epidemiology, and Clinical Manifestations of Celiac Disease in Adults.*Waltham, MA: *UpToDate*; 2008

Barton, S.H., Kelly, D.G., Murray, J.A. *Nutritional Deficiencies in Celiac Disease. Gastroenterol Clin North Am.* 2007;36:93-108.

Fasano, A., Berti, I., Gerarduzzi, T., et al. *Prevalence of Celiac Disease in at-risk and not-at-risk Groups in the United States: A Large Multicenter Study. Arch Intern Med.* 2003;163(3):286–292.

Rostom, A., Murray, J.A., Kagnoff, M.F. *American Gastroenterological Association (AGA) Institute Technical Review on the Diagnosis and Management of Celiac Disease. Gastroenterology.* 2006;131(6):1981–2002.

Greco, L., Romino, R., Coto, I., et al. *The First Large Population based Twin Study of Coeliac Disease. Gut.* 2002;50(5):624–628.

Collin, P., Kaukinen , K., Välimäki, M., Salmi, J. *Endocrinological Disorders and Celiac Disease. Endocr Rev.* 2002;23(4):464–483.

Cataldo, F., Marino, V., Ventura, A., Bottaro, G., Corazza, G.R. *Prevalence and Clinical Features of Selective Immunoglobulin a Deficiency in Coeliac Disease: An Italian Multicentre Study.* Italian Society of Paediatric Gastroenterology and Hepatology (SIGEP) and "Club del Tenue" Working Groups on Coeliac Disease. *Gut.* 1998;42(3):362–365.

National Institutes of Health Consensus Development Conference Statement on Celiac Disease, June 28–30, 2004. *Gastroenterology.* 2005;128(4 suppl 1) :S1–S9.

Treem, W.R. *Emerging Concepts in Celiac Disease. Curr Opin Pediatr.* 2004; 16(5): 552–559.

Accomando, S., Cataldo, F. *The Global Village of Celiac Disease. Dig Liver Dis.* 2004;36(7): 492–498.

Tommasini, A., Not, T., Kiren, V., et al. *Mass Screening for Coeliac Disease using Antihuman Transglutaminase Antibody Assay. Arch Dis Child.* 2004;89(6):512–515.

Hill, I.D. *What are the Sensitivity and Specificity of Serologic Tests for Celiac Disease*? Do sensitivity and specificity vary in different populations? *Gastroenterology*. 2005;128(4 suppl 1):S25–S32.

Lewis, N.R., Scott B.B. *Systematic Review: the use of Serology to Exclude or Diagnose Coeliac Disease* (a comparison of the endomysial and tissue transglutaminase antibody tests). *Aliment Pharmacol Ther*. 2006;24(1): 47–54.

AGA Institute. AGA Institute Medical Position Statement on the Diagnosis and Management of Celiac Disease. *Gastroenterology*. 2006;131(6):1977–1980.

Fotoulaki, M., Nousia-Arvanitakis S, Augoustidou-Savvopoulou P, Kanakoudi-Tsakalides F, Zaramboukas T, Vlachonikolis J. *Clinical Application of Immunological Markers as Monitoring Tests in Celiac Disease. Dig Dis Sci.* 1999;44:2133–8.

Shield, J., Mullin, M.C. *Patient Education Materials. Supplement to the Manual of Clinical Diatetics*. 3rd ed. Chicago, Ill.: American Dietetic Association; 2001.

Aggarwal, S., Lebwohl, B., Green, P.H., Therap Adv Gastroenterol. 2012 Jan;5(1): 37-47. *Screening for Celiac Disease in Average-risk and High-risk Populations*. PMID: 22282707.

OO

Green Tea Against Malignancies

Charli Kaushal

Abstract

The concept of prevention of cancer using naturally occurring substances that could be included in the diet consumed by the human population is gaining increasing attention. Tea, next to water, is the most popularly consumed beverage in the world and it is grown in about 30 countries. Various plants and their constituents have show their beneficial therapeutic effect including anti-oxidant, anti-inflammatory, anticancer and immunomodulatory effect and green tea is one of them. Green tea, produced from the plant of ***Camellia sinensis****. Green tea originated in China, China has hundreds of different types of green teas. Other producers of green tea include India, Indonesia, Korea, Nepal, Sri Lanka, Tiwan, and Vietnam. The anti-carcinogenic and anti-mutagenic activities of green tea reduce the prevalence of cancer and even provide protection. The pharmacological action of green tea are mainly attributed to polyphenols. Polyphenol is most important chemical compound that includes epigalocatechin-3-gallate (EGCG), epicatechin, epicatechin-3-gallate, epigallocatechin. It also contain other agents that have chemopreventive activities. These include caffeine, flavaniods, phenolic acid as well as the alkaloids theobromine and theophylline. Cancer chemopreventive properties of green tea are mediated by EGCG that includes apoptosis promotes cell growth arrest, by altering the expression of cell cycle regulatory proteins, activating killer caspases, and suppressing kappa nuclear factor-B activation. Green tea is also regulates and promotes IL-23*

Dependent DNA repair and stimulates cytotoxic T-cells activities in a tumor microenvironment. It is also blockcarcinogenesis by modulating signal transduction pathways involve in the cell proliferation, transformation, inflammation and metastasis. The review is highlight the chemistry of green tea, Its anti-oxidant potential and mode of action against various cancer cell lines that showed its potential as a chemopreventive agent against colon, skin, lung, prostate and breast cancer.

Keywords: *Green Tea, Nutritional Support, Epigalocatechin-3-gallate (EGCG), Malignancies.*

Introduction

Various plants and their constituents have show their beneficial therapeutic effect including anti-oxidant, anti-inflammatory, anticancer and immunomodulatory effect and green tea is one of them (Miller et al., 2004; Huffman, 2003; Parab et al., 2003; Fong, 2002). Green tea is a type of tea made solely with the leaves of ***Camellia sinensis*** that has under gone minimum oxidation during processing (Nagao T et al., 2007). Green tea originated in China for the medicinal purpose, and its first recorded use was 4,000 years ago. By the third century, it became a daily drink and cultivation and processing began. Today, China has hundreds of different types of green teas. Other producers of green tea include India, Indonesia, Korea, Nepal, Sri Lanka, Taiwan and Vitenam.

The term "green tea" refers to the product manufactured from fresh tea leaves by steaming and drying at elevated temperatures with the precaution to avoid oxidation of the polyphenolic components (Chow and Kramer, 1990). Green tea made from unfermented leaves, Oolong tea is made of semifermented while black tea is made of nearly fully fermented product and their processing involves the oxidation reaction which is catalyzed by Polyphenol oxidase (PPO) (Wargovich MJ et al., 2001). PPO is mainly responsible for catalyzed oxidation of tea leaf catechins (Wilson and Clifford, 1992).

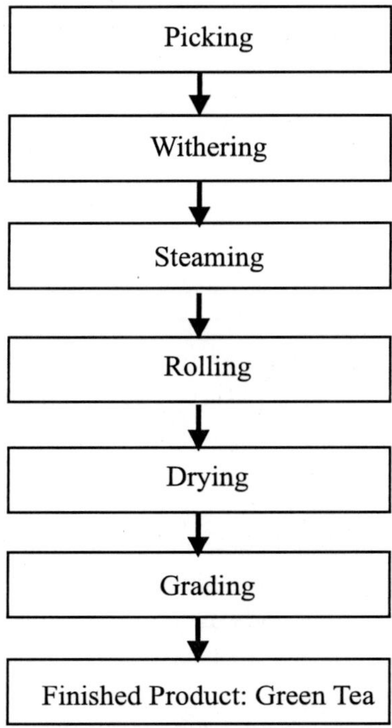

Figure 20.1: *Processing Flow Chart of Green Tea*

With green tea, the freshly picked leaves are steamed, rolled and dried so that the green color of the chlorophyll is retained. The fermentation process is omitted.

TABLE 20.1

Consumption of Tea

Type of Tea	*% Share in Consumption*	*Countries*
Black tea	78%	Western countries and some Asian countries
Green tea	20%	Asian countries mainly Japan, China, Korea & India
Oolong tea	2%	Southeastern China

Sources: Hara (2001); Katiyar and Mukhtar (1996).

Botanical Classification

Kingdom Plantae – Plants
Subkingdom Tracheobionta – Vascular plants
Superdivision Spermatophyta – Seed plants
Division Magnoliophyta – Flowering plants
Class Magnoliopsida – Dicotyledons
Subclass Dilleniidae
Order Theales
Family Theaceae – Tea family
Genus Camellia L. – camellia
Species Camellia sinensis (L.) – tea

Figure 20.2

Chemical Constituents

The active constituents in Green Tea are, catechin polyphenols, epigallocatechin gallate (EGCG),vitamin & mineral, amino acid & other nitrogenous compound, inorganic elements and carbohydrates etc. Green tea ingredients are extremely complex. It contains as many as 200 bioactive compounds. (Westerterp-Plantenga MS et al., 2005)

Polyphenols

The largest and most important chemical compound is **polyphenols**. The primary function of polyphenol is to scavenge the free radicals that occur in the body. Free radicals occur naturally in the body causing the ageing process and numerous health hazards like cancer and heart diseases (Suzuki Y et al., 2004). Polyphenols contain flavonoids-an important class of antioxidants, antiinflammatory, and anti-carcinogenic properties and such health benefits of tea are mainly attributed to its polyphenols contents (Higdon and Frei, 2003; Katiyar et al., 2001). Polyphenol contains flavonoids (catechin, epicatechin, epicatechin gallate, epigallocatechin gallate.) (Li L, Wang Z et al.,2005; Hodgson JM et al., 1999; Lambert Jd et al., 2003). According to research, the strong antioxidant property of green tea are mainly due to the higher concentration of EGCG.

Green Tea Polyphenols

- decrease Insulin-like Growth Factor-1 (IGF-1), while increasing levels of IGF binding protein-3, which binds IGF-1, further diminishing its activity. (Increased levels of IGF-1 are associated not only with prostate cancer, but cancers of the breast, lung and colon.)
- inhibit key cell survival proteins, promoting apoptosis or programmed cell death in cancer cells.
- reduce the expression of several compounds (urokinase plasminogen activator and matrix metalloproteinase 2 and 9) involved in the metastasis and spread of cancer cells.
- reduce the amount of Vascular Endothelial Growth Factor (VEGF), which develops new blood vessels to carry nutrients to developing tumors (Adhami VM et al., 2004, Chen D et al., 2004)

Catechins

Catechins were discovered in 1970 when a study was made in Shizuoka, Japan. It has been discovered that people living in Shizuoka Prefecture has much lower rates of cancer. The four main catechins are: shown in Figure Epigallocatechin gallate EGCG is a powerful anti-oxidant, inhibiting the growth of cancer cells, kills cancer cells without any harming healthy tissue and also effective in lowering LDL cholesterol levels, and inhibiting the abnormal formation of blood clots (Nagata T;1990). Current research has shown EGCG to be as much as 100 times more potent than vitamin C and 25 time more potent than vitamin E in antioxidant power. Antimicrobial properties in green tea are also attributed to EGCG, repress the growth of bacteria such as helicobacter pylori, E. coli, and *Staphylococcus aureus*. The active catechins and their respective concentrations in green tea infusions are listed in the table below.

TABLE 20.2

Catechin Concentrations of Green Tea Infusion

Catechin in Green Tea Infussion	*Catechin Concentration (mg/l)*	*Catechin Concentration (mg/8 fl oz)*
Epigallocatechin-3-gallate (EGCG)	117–442	25–106
Epigallocatechin (EGC)	203–471	49–113
Epicatechin-3-gallate (ECG)	17–150	4–36
Epicatechin (EC)	25–81	6–19

mg = milligram; L = liter; fl oz = fluid ounce.
(Reto M et al., 2007)

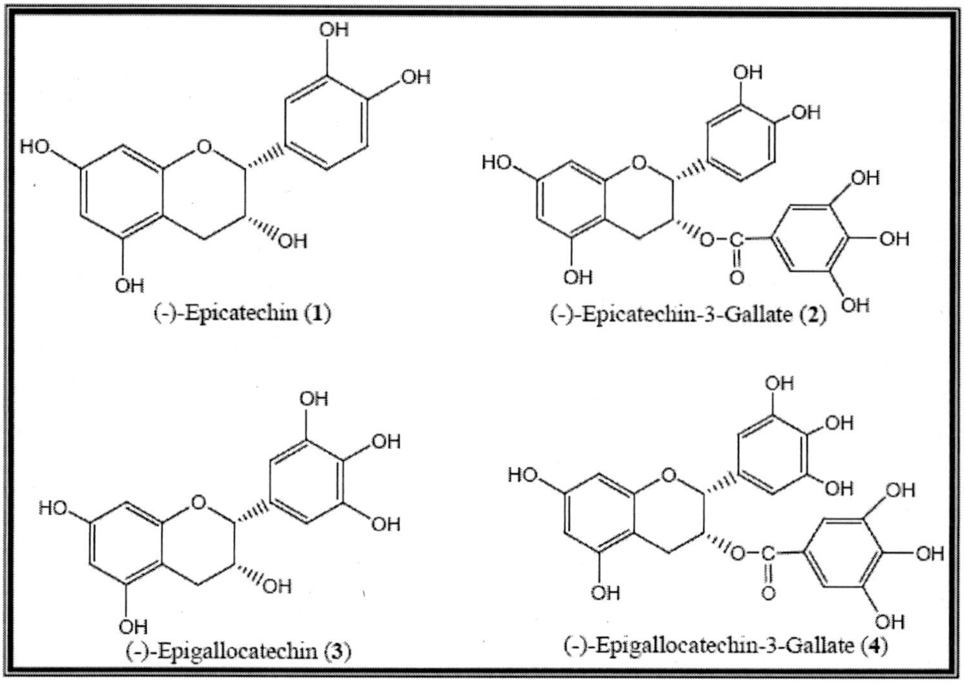

(-)-Epicatechin (1)

(-)-Epicatechin-3-Gallate (2)

(-)-Epigallocatechin (3)

(-)-Epigallocatechin-3-Gallate (4)

Figure 20.3: *Structure of Epitcatechin, (1)Epitcatechin-3- gallate, (2) Epigallocatechin,(3) Epigallocatechin-3-gallate*

Cancer.2001;92(4):600-604. Cancer.2001;92(4):600-604.

Caffeine

Green tea caffeine content depends largely on the size and type of the tea leaf used. That is, the larger the leaf, the less caffeine. Also, the more steeping, the less caffeine you have. Leaves of green tea plants grown under shade will develop more caffeine in them than those grown in the open. The highest green tea caffeine levels are found in the top two leaves and a bud, whereas green tea leaves picked from the lower parts of the bush, such as bancha green tea, have much less caffeine content than those picked from the higher parts. Green tea contains 32 mg caffeine per grams of dry tea. Green tea contains alkaloids known as methylxanthines such as caffeine, theobromine and theophylline. It acts as natural pesticide, protecting plants against certain insects feeding on them. Graham found that fresh leaves contain, on average, 3% to 4% of caffeine and very little amounts of the methylxanthines.

Vitamins and Minerals

Green tea contains several vitamins B and vitamin C, include 6% to 8% of minerals such as aluminium, fluoride and manganese, also contains organic acids such as Gallic and quinic acids, and 10% to 15% of carbohydrate and small amount of volatiles.

Theanine

Dried tea extract contains 4% to 6% of **theanine**; gives the characteristics flavor. Taste of Catechins and caffeine bitter and astringent, but the tastes of theanine sweet and fresh. Theanine stimulates alpha brain waves, calms the body and promotes relaxed awareness causing agents.

Carbohydrate

It contains 40% carbohydrate. Carotenoids, tocopherols, ascorbic acid (vitamin C), minerals such as chromium, manganese, selenium or zinc, and certain phytochemical compounds. (Pianetti S,et al., 2002; Rowe CA,et al., 2007; Setiawan VW,et al., 2001; Shankar S,et al., 2008; Sasazuki S,et al., 2000; Choan E,et al., 2005)

Malignancies and Green Tea

Cancer is uncontrolled cell division where cells aggregate and formation of tumors take place. In the initial step of carcinogenesis metabolic activation of chemical carcinogens by the P-450-dependent biotransformation reaction take place. (Mukhtar et al., 1991).In a living system life span of both normal and cancer cells is affected by the rate of apoptosis ("programmed" cell death). Much of the cancer chemopreventive properties of green tea are mediated by EGCG. EGCG modulates the signal transduction pathway which involve in cell proliferation, transformation, inflammation, apoptosis, and metastasis (Khan et al., 2006; Na and Surh, 2006; Gupta et al., 2004).

Colon Cancer

Several studies have shown that green tea helps to protect against colon cancer. EGCG treatment to colon cancer cells results in strong activation of AMPK, an inhibition of COX-2 expression. Increased expression of COX 2 appears to play a serious role in the development of colorectal cancer (Shimizu et al., 2005). The effect of EGCG on COX-2 expression resulted in decreased COX-2 promoter activity via inhibition of nuclear factor kappa-B activation (Lambert et al., 2006; Balavenkatraman et al., 2006).

Skin Cancer

Polyphenols from green tea have anti-inflammatory & anti cancer property that help in prevention of onset & growth of skin tumor. Ultraviolet B (UVB) radiation present in the solar spectrum is one of the major risk factors for skin cancer in humans (Elmets,1991). Infusion of green tea extracts as a source of drinking water

(125%, w/v) to mice afforded protection against UVB radiation-induced intensity of red color and area of skin lesions, as well as UVB radiation-induced tumor initiation and tumor promotion (Wang et al., 1992a; b).

Lung Cancer

Lung cancer, also known as bronchogenic cancer. Treatment of lung cancer depends upon histopathologic groupings into Small Cell Lung Carcinoma (SCLC) and Non-Small Cell Lung Carcinoma (NSCLC). Sadava et al. (2007) observed that incubation of human SCLC cells in EGCG for 24 h resulted in 50–60% reduced telomerase activity and reduction in activities of caspases 3(50%) and caspases 9(70%) but caspase 8 and DNA fragmentation remain unaffected with EGCG treatment. It has the capacity to block the cell-cycle in S phase indicating the potential use of EGCG, and possibly green tea, in treating SCLC.

Prostate Cancer

Prostate cancer is the most common male cancer. The treatment with tea ingredients results in (i) significant inhibition in growth of implanted prostate tumors, (ii) reduction in the level of serum prostate specific antigen, (iii) induction of apoptosis accompanied with upregulation in Bax and decrease in Bcl-2 proteins and (iv) decrease in the level of VEGF protein (Siddique et al., 2006). EGCG inhibits COX-2 without affecting COX-1 expression at both the mRNA and protein levels, in androgen-sensitive LNCaP and androgen-insensitive PC-3 human prostate carcinoma cells (Paschka et al., 1998).

Breast Cancer

EGCG suppressed cell viability and induced apoptosis by the down-regulation of telomerase and inhibited angiogenesis by reducing the expression of Vascular Endothelial Growth Factor (VEGF) in a dose-dependent (Mittal et al., 2004; Sartippour et al., 2002). In vitro, epigallocatechin, another major catechin in green tea, also has strong effects in inducing apoptosis and inhibiting growth of breast cancer cells (Stuart et al., 2007; Zhao et al., 2006; Vergote et al., 2002).

Mechanism of Action

Multiple mechanisms are involved, including induction of apoptosis, cell cycle arrest down regulation of telomerase, inhibition of vascular endothelial growth factor, and suppression of aromatase activity. The protective effects of green tea polyphenols have been attributed to the inhibition of enzymes such as the cytochromes P450, which are involved in the bio-activation of carcinogens.The green tea polyphenols stimulate the transcription of Phase II detoxifying enzymes through Antioxidant-Responsive Element (ARE). Mukhtar and Ahmad, (1999) suggested that green tea polyphenol treatment results in significant activation of MAPK, extracellular, signal regulated kinase-2 (ERK2), as well as JNK1 and an increase in the mRNA levels of early response genes c-*jun* and c-*fas*. Jankun et al., (1997) also highlighted the anti-cancer activity of EGCG and associated it with the inhibition of urokinase, is one of the most frequently expressed enzymes in human cancers.

TABLE 20.3

World Tea Production 2006–2008
(in Metric Tons and as % of Total World Production)

Country	*2006 Quantity*	%	*2007 Quantity*	%	*2008 Quantity*	%
China	1,047,350	29%	1,183,000	30%	1,257,385	33%
India	928,00	25%	949,22	24%	805,180	21%
Kenya	310,580	9%	369,600	9%	345,800	9%
SriLanka	310,800	9%	305,220	8%	318,470	8%
Turkey	201,900	6%	206,160	5%	210,000	5%
Vietnam	151,500	4%	164,000	4%	174,900	4%
Indonesia	146,850	4%	150,200	4%	150,850	4%
Japan	91,800	3%	94,100	2%	94,100	2%
Argentina	72,100	2%	76,000	2%	76,000	2%
Iran	59,200	2%	60,000	2%	60,000	2%
Bangladesh	58,000	2%	58,500	1%	59,000	2%
Malawi	45,500	1%	46,000	1%	46,000	1%
Uganda	34,300	1%	44,900	1%	42,800	1%
World Total	3,649,170	100%	3,902,880	100%	3,845,700	100%

Source: Data from de FAOSTAT, except for Turkey 2008, which is estimated (FAO database, accessed April 16, 2010)

Precaution

In Anemic

Due to high tannin content, green tea prevents iron absorption.

In Expecting Mothers

High amount of EGCG acts like the anticancer drugs, which inhibit dihydrofolate enzyme which the cancer cells utilize to grow & proliferate. The embryo also utilizes the dihydrofolate enzyme for cell proliferation and differentiation. (Mukai T, et al., 1992)

Conclusions

The health benefits of green tea are often attributed to polyphenols especially EGCG, recognized as an effective chemopreventive agent that holds the ability to scavenge free radicals and is certainly helpful against various cancers. In conclusion the consumption of green tea regularly or its possible inclusion in diet-based therapies could yield certain health benefits and potential defense against malignancies.

References

Adhami, V,M,, Siddiqui, I.A., Ahmad, N., Gupta, S., Mukhtar, H. *Oral Consumption of Green Tea Polyphenols Inhibits Insulin-like Growth Factor-I-induced Signaling in an Autochthonous Mouse Model of Prostate Cancer*. Cancer Res. 2004 Dec 1;64(23):8715-22. 2004. PMID:15574782

Balavenkatraman, K.K., Jandt, E., Friedrich, K., Kautenburger, T., Pool-Zobel, B.L., Ostman, A. and B¨ohmer, F. D. (2006). *DEP-1 Protein Tyrosine Phosphatase Inhibits Proliferation and Migration of Colon Carcinoma Cells and is Upregulated by Protective Nutrients*. Oncogene. 25:6319–24.

Chen, D., Daniel, K.G., Kuhn, D.J., Kazi, A., Bhuiyan, M., Li, L., Wang, Z., Wan, S.B., Lam, W.H., Chan, T.H., Dou, Q.P. *Green Tea and Tea Polyphenols in Cancer Prevention*. Front Biosci. 2004 Sep 01; 9:2618-31. 2004. PMID:15358585

Choan, E., Segal, R, Jonker D, et al. *A Prospective Clinical Trial of Green Tea for Hormone Refractory Prostate Cancer: An Evaluation of the Complementary/Alternative Therapy Approach*. Urol Oncol 2005; 23(2):108-113.

Chow, K. and Kramer, I. (1990). *All the tea in China*. San Francisco: China Books and Periodicals

Elmets, C.A. (1991). *Cutaneous Photocarcinogenesis*. In: Pharmacology of the Skin. pp. 389–416. Mukhtar H(ed.). CRC Press, Boca Raton, Florida.

Fong, H.H. (2002). *Integration of Herbal Medicine into Modern Medical Practices: Issues and Prospects*. Integr. Cancer. Ther. 1:287–93.

Gupta, S., Hastak, K., Afaq, F., Ahmad, N. and Mukhtar, H. (2004). *Essential Role of Caspases in Epigallocatechin-3-gallate-mediated Inhibition of Nuclear Factor Kappa B and Induction of Apoptosis*. Oncogene. 23:2507–2522.

Higdon, J.V. and Frei, B. (2003). *Tea Catechins and Polyphenols: Health Effects, Metabolism, and Antioxidant Functions*. Crit. Rev. Food. Sci. Nutr. 43:89–143.

Huffman, M.A. (2003). *Animal Self-medication and Ethno-medicine: Exploration and Exploitation of the Medicinal Properties of Plants*. Proc. Nutr. Soc. 62:371–81

Jankun, J., Selman, S.H., Swiercz, R. and Skrzypczak-Jankun, E. (1997). *Why Drinking Green Tea could Prevent Cancer*. Nature 387:561.

Katiyar, S. K., Bergamo, B. M., Vyalil, P.K. and Elmets, C. A. (2001). *Green Tea Polyphenols: DNA Photodamage and Photoimmunology*. J. Photochem. Photobiol. B. 65:109–14.

Khan, N., Afaq, F., Saleem, M., Ahmad, N. and Mukhtar, H. (2006). *Targeting Multiple Signaling Pathways by Green Tea Polyphenol* (")-epigallocatechin- 3-gallate. Cancer. Res. 66:2500–2505

Lambert, J.D., Lee, M. J., Diamond, L., Ju, J., Hong, J., Bose, M., Newmark, H.L.,Yang, C. S. (2006). *Dose-dependent levels of epigallocatechin-3-gallate in human colon cancer cells and mouse plasma and tissues*. Drug. Metab. Dispos. 34(1):8–11.

Miller, K.L., Liebowitz, R.S. and Newby, L.K. (2004). *Complementary and Alternative Medicine in Cardiovascular Disease: A Review of Biologically based Approaches*. Am. Heart. J. 147:401–11.

Mittal, A., Pate, M., Wylie, R., Tollefsbol, T. and Katiyar, S. (2004). *EGCG down-regulates Telomerase in Human Breast Carcinoma MCF-7 cells, leading to Suppression of Cell Viability and Induction of Apoptosis*. Int. J. Oncol. 24:703–10.

Mukai, T., Horie, H., and Goto, T., *Differences in Free Amino Acids and Total Nitrogen Content among Various Prices of Green Tea,* Tea Res J 1992; 76: 45

Mukhtar, H., Agarwal, R. and Bickers, D.R. (1991). *Cutaneous Metabolism of Xenobiotics and Steroid Hormones*. In: Pharmacology of the Skin. pp. 89–110. Mukhtar, H. (ed.). CRC Press, CRC Press, Boca Raton, Florida

Na, H.K. and Surh, Y.J. 2006. *Intracellular Signaling Network as a Prime Chemopreventive Target of (")-epigallocatechin gallate*. Mol. Nutr. Food. Res. 50:152–9.

Nagata, T. *New Analytical Methods Studying Quality Components Tea* Res J, 1990;72:53

Nagao, T., Hase, T., Tokimitsu I. *A Green Tea extract High in Catechins reduces Body Fat and Cardiovascular Risks in Humans*. Obesity (Silver Spring). 2007;15(6):1473-83

Parab, S., Kulkarni, R. and Thatte, U. (2003). *Heavy Metals in dherbalT Medicines*. Indian. J. Gastroenterol. 22:111–2.

Paschka, A.G., Butler, R.,Young, C.Y. (1998). *Induction of Apoptosis in Prostate Cancer Cell Lines by the Green Tea Component*, (-)-epigallocatechin-3-gallate. Cancer Lett. 130(1–2):1–7

Pianetti, S., Guo, S., Kavanagh, K.T., Sonenshein, G.E. *Green Tea Polyphenol Epigallocatechin-3 Gallate Inhibits Her-2/neu Signaling, Proliferation, and Transformed Phenotype of Breast Cancer Cells*. Cancer Res. 2002;62(3):652-655.

Reto, M., Figueira, M.E., Filipe, H.M., Almeida, C.M. *Chemical Composition of Green Tea (Camellia Sinensis) Infusions commercialized in Portugal*. Plant Foods for Human Nutrition 2007; 62(4):139–144

Rowe, C.A., Nantz, M.P., Bukowski, J.F., Percival, S.S. *Specific Formulation of Camellia Sinensis prevents Cold and Flu Symptoms and Enhances Gammadelta T Cell Function: A Randomized, Double-blind, Placebo-controlled Study*. J Am Coll Nutr. 2007;26(5):445-52.

Sadava, D., Whitlock, E. and Kane, S.E. (2007). The Green Tea Polyphenol, Epigallocatechin-3-gallate Inhibits Telomerase and induces Apoptosis in Drugresistant Lung Cancer Cells. Biochem. Biophys. Res. Commun. 360(1):233–7.

Sartippour, M.R., Shao, Z.M., Heber, D., Beatty, P., Zhang, L., Liu, C., Ellis, L., Liu, W., Go, V.L. and Brooks, M. N. (2002). *Green Tea inhibits Vascular Endothelial Growth Factor (VEGF) induction in Human Breast Cancer Cells*. J. Nutr. 132:2307–11.

Sasazuki, S., Kodama, H., Yoshimasu, K. et al. *Relation between Green Tea Consumption and the Severity of Coronary Atherosclerosis among Japanese Men and Women*. Ann Epidemiol. 2000;10:401-408.

Setiawan, V.W., Zhang, Z.F., YU, G.P., et al. *Protective Effect of Green Tea on the Risks of Chronic Gastritis and Stomach Cancer*. Int J Cancer. 2001;92(4):600-604

Siddiqui, I.A., Zaman, N., Aziz, M.H., Reagan-Shaw, S.R., Sarfaraz, S., Adhami, V.M., Ahmad, N., Raisuddin, S. and Mukhtar, H. (2006). *Inhibition of CWR22Rnu1 Tumor Growth and PSA Secretion in Athymic Nude Mice by Green and Black Teas*. Carcinogenesis. 27(4):833–9.

Shankar, S., Ganapathy, S., Hingorani, S.R., Srivastava, R.K.. *EGCG inhibits Growth, Invasion, Angiogenesis and Metastasis of Pancreatic Cancer*. Front Biosci. 2008; 13:440-52.

Shimizu, M., Deguchi, A., Joe, A.K., Mckoy, J.F., Moriwaki, H., Weinstein, I.B. (2005). *EGCG inhibits Activation of HER3 and Expression of Cyclooxygenase-2 in Human Colon Cancer Cells*. J. Exp. Ther. Oncol. 5(1):69–78.

Stuart, E.C., Scandlyn, M.J., and Rosengren, R.J. (2007). *Role of Epigallocatechin Gallate (EGCG) in the Treatment of Breast and Prostate Cancer.* Life Sci. 79:2329–36.

Suzuki, Y., Tsubono, Y., Nakaya, N., Suzuki, Y., Koizumi, Y., Tsuji, I. *Green Tea and the Risk of Breast Cancer: Pooled Analysis of Two Prospective Studies in Japan.* Br J Cancer. Apr 5, 2004; 90(7)1361-1363.

Vergote, D., Cren-Olive, C., Chopin, V., Toillon, R.A., Rolando, C., Hondermarck, H. and Le Bourhis, X. (2002). *(-)-Epigallocatechin (EGC) of Green Tea induces Apoptosis of Human Breast Cancer Cells but not of their Normal Counterparts.* Breast Cancer Res. Treat. 76:195–201.

Wargovich, M.J., Woods, C., Hollis, D.M., Zander, M.E. Herbals, *Cancer Prevention and Health.* [Review]. J Nutr. 2001; 131(11 Suppl):3034S-3036S

Wang, Z.Y., Khan, W.A., Bickers, D.R. and Mukhtar, H. (1989b). *Protection against Polycyclic Aromatic Hydrocarbon-induced Skin Tumor Initiation in Mice by Green Tea Polyphenols.* Carcinogenesis 10:411–15

Wang, Z.Y., Huang, M.T., Ferraro, T., Wong, C.Q., Lou, Y.R., Latropoulos, M., Yang, C.S. and Conney, A.H. (1992a). *Inhibitory Effect of Green Tea in the Drinking Water on Tumorigenesis by Ultraviolet Light and 12-Otetradecanoylphorbol- 13-acetate in the Skin of SKH-1 Trice.* Cancer Res. 52:1162–70.

Wang, Z.Y., Huang, M.T., Ho, C.T., Chang, R., Ma, W., Ferraro, T., Reuhl, K.R., Yang, C.S. and Conney, A.H. (1992b). *Inhibitory Effect of Green Tea on the Growth of established Skin Papillomas in Mice.* Cancer Res. 52:6657– 65.

Westerterp-Plantenga, M.S., Lejeune, M.P., Kovacs, E.M. *Body Weight and Weight Maintenance in Relation to Habitual Caffeine Intake and Green Tea.* Obes Res Jul2005;13(7):1195-1204

Zhao, X., Tian, H., Ma, X. and Li, L. (2006). *Epigallocatechin Gallate, the main Ingredient of Green Tea induces Apoptosis in Breast Cancer Cells.* Front Biosci. 111:2428–33.

21

Berseem Clover (Trifolium Alexandrinum) as a Lmmuno Nutrient

Pant P, Rustagi S, Yadav S. and Mathur S

Abstract

Berseem (Trifolium alexandrinum) clover is an annual pasture legume originating in eastern Mediterranean regions. It is widely grown as a high-quality forage conservation crop, and has been introduced to many other countries, such as India, Pakistan, South Africa, USA and Australia & is also primarily used for fodder conservation purposes. Berseem clover is also known as Egyptian and Alexandria clover. Berseem clover grows best on fertile, medium to heavy textured soils of mildly acidic to neutral pH. Sowing is to be done in mid October. The crop can be harvested in 45 days after sowing and subsequent cuts can be taken after every 30 days. It has moderate tolerance of salinity, and will tolerate short periods of water logging. Berseem clover is generally only sown as a one-year fodder conservation crop, as it produces soft seed, which is highly susceptible to loss via false breaks (out of season rainfall with no follow-up rain) in summer and early autumn. As false autumn breaks are common in areas where berseem clover is sown, second year stands are usually very sparse. Berseem can grow 60–80 cm tall. It has a shallow taproot. Leaflets are commonly 4–5 cm long and 2–3 cm wide. Flowers are round, white and approximately 2 cm in diameter. Seeds are yellow, with approximately 400,000 seeds/kg. It has good forage quality, digestibility and high palatability for cattle. In fact, it is known as milk multiplier of cattle's. Being a leguminous crop it fixes atmospheric nitrogen and thus is also good for soil fertility (Malaviya, et al., 2010).Berseem fresh leaves contain 18-23% protein and is very rich in minerals and vitamins. Berseem is successfully used for animals but not at all for humans. They sell it to market as animal fodder at low price. Farmers can process Berseem extract in powder form, fit for human consumption. With berseem an increase in serum pro inflammatory cytokines (TNFá, IFNã), and anti-inflammatory mediators (IL-10) but a fall in(IL-4) levels was observed. They are found to be immunostimulatory nutrient as they increased the levels of CD8 and CD4 simultaneously.

Keywords: *Berseem, Immunostimulatory Agent, Cytokines, CD8, CD4.*

Introduction

An Egyptian clover (Trifolium Alexandrinum) extensively cultivated as a forage plant and soil- renewing crop in the alkaline soils of the Nile valley, and now introduced into the southwestern United States. It is more succulent than other clovers or alfalfa. called also Egyptian clover grown in India for animal fodder. Fresh leaves contain 18-23% protein. It is a winter forage legume in India, Pakistan, Turkey and Egypt. Berseem is the most important winter season legume cultivated in an area of around 2 million hectares in India, particularly, central, north, north-west and north-eastern parts. Sowing is to be done in mid October. The crop can be harvested in 45 days after sowing and subsequent cuts can be taken after every 30 days. The crop gives 6 to 7 cuts during November – May and yield ranges from 100-120 t/ha. It has good forage quality (20% Crude protein), digestibility (up to 65%) and high palatability for cattle. In fact, it is known as milk multiplier of cattle's. Being a leguminous crop it fixes atmospheric nitrogen and thus is also good for soil fertility (Malaviya, et al., 2010). Berseem (Trifolium alexandrinum) leaves in diet as immuno-nutrient; cytokine and T-cell subpopulation response in malnutrition.

Nutritional Status in INDIA

India has >50% under/malnourished mothers and maternal anemia in >85% (Agarwal et al. ICMR; 2006); 30% births are Low Birth Weight; 47% children < 3 years are malnourished and of these 2.8% are having acute severe malnutrition needing hospital admission, A situation of Endemic Malnutrition in between two National Family Health Surveys (1998-99) to (2005-6), there was only 1% reduction in malnutrition (47% to 46%); while severe malnutrition reduced from 20% to 18% only.

Choice of Food Source(s) to Combat Malnutrition

Plant foods like leaves contain almost all of the mineral and organic nutrients established as essential for human nutrition, as well as a number of unique organic phytochemicals that have been linked to the promotion of good health. Composition of most of the leaf proteins is similar, which justifies their use in our mixed diet to meet the needs of micronutients and protein. Because the concentrations of many of these dietary constituents are often low in edible plant sources, "Leaf protein concentrates" are developed.

Development of Leaf Protein Concentrate

The protein content of common Indian leafy vegetables (100g eatable portion) is: Amaranth 2.5 g; Mustard-Indian 2.7g; Turnip 1.5g; Broccoli 1.8g; Cauliflower leaves 2 g; and Moringa oleifera (drum sticks) 100 g dried leaves give 29g protein. However the micronutrient content is similar in these leaves. Berseem (Trifolium alexandrinum) has been used for animal fodder, its fresh leaves contain 18-23%

protein and is very rich in mineral and vitamins. Thus Berseem was chosen to develop Leaf Protein Concentrate (LPC) by ultrafiltration and acid thermocoagulation. Generally, LPC is produced by pulping leaves and pressing the juice out, heating the juice to coagulate the protein, and filtering the protein out and drying it.

Barseem Crop -> Trimming -> Washing -> Chopping -> Extraction -> Drying -> Powder -> Storage

Immunonutrient Value of Barseem

In one study carried out by Pooja Dewan, I.R. Kaur*, M.M.A. Faridi & K.N. Agarwal** Eighty moderately and severely malnourished children, 1-5 yr of age, received the WHO recommended diet for severe malnutrition, modified according to local dietary habits, containing in addition either curd or micronutrient-rich leaf protein concentrate, for a period of 15 days. Cytokine levels [tumour necrosis factor α (TNFα), interferon γ (IFNγ), interleukin-10 (IL-10), interleukin-4 (IL-4)] were measured before and after dietary rehabilitation.

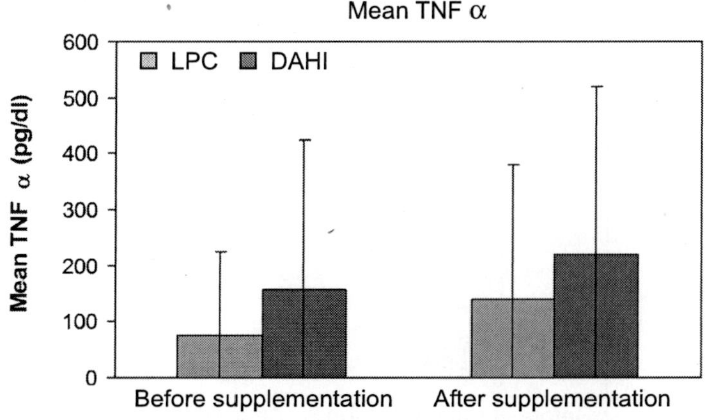

Figure 21.1: *TNF α(RISE-LPC 90%& FM 39%)*

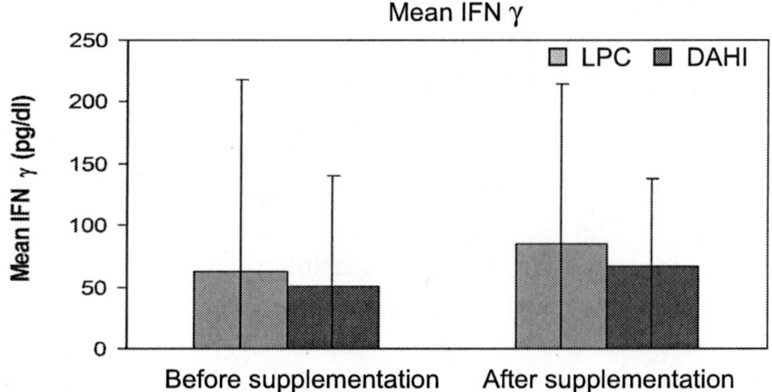

Figure 21.2: *IFN γ(RISE LPC 36% & FM 32%)*

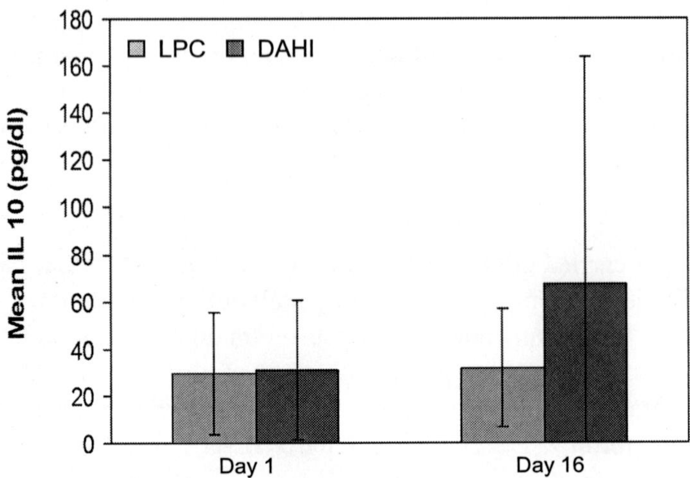

Figure 21.3: *(IL-10(RISE-LPC 8% and FM 118%)*

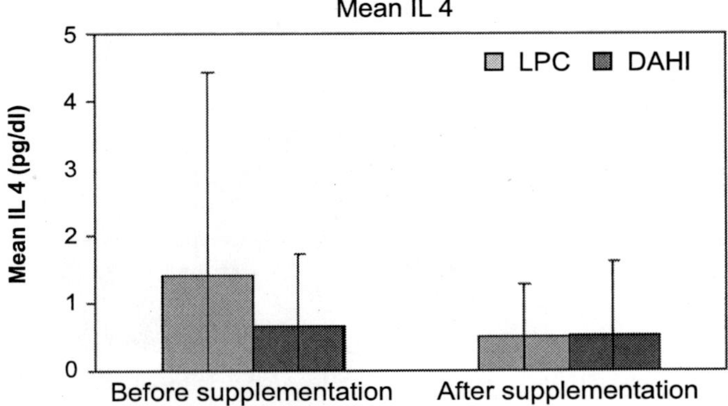

Figure 21.4: *IL-4 Decreased LPC 67% & FM 23%)*

In other study, it was shown that consumption of curd and Leaf Protein Concentrate (LPC) derived from the leaves of Trifolium alexandrinum in malnourished children significantly improved the weight, haemoglobin, and T-lymphocyte ratio (CD4/ CD8). Other studies have also shown that lactic-acid bacteria can influence cytokine production12-14. Though curd is an integral constituent of the Indian diet, leaf protein concentrate derived from olive leaf (Olea europeal), berseem (Trifolium alexandrinum) and tulsi (Oscimum sanctum) may be a cheap alternative as these green leaves are rich sources of micronutrients like carotenoids, zinc, and vitamin A, which have been proved to have good immunomodulatory action. As a dietary supplementation, the levels of pro-inflammatory cytokines (TNFα, IFNγ) increases. This was seen despite all children showing clinical improvement from infections. Thus, the initial cytokine response in malnourished children may have been suppressed. Solis et al. AJCN;1997, also reported an increase in the pro-inflammatory cytokine IFNγ in malnourished children following nutritional rehabilitation. There was an increase in IL-10 production may be to counter the damaging effects of increased pro-inflammatory cytokines (TNFα, IFNγ). The fall in IL-4 levels is explainable as it is solely produced by the lymphocytes, and whose function is seriously affected. PEM has been known to be associated with a decrease in absolute T lymphocyte count and a reduction in CD4: CD8 counts. The rise in the CD4: CD8 ratio after nutrition intervention supported immunoenhancing effects.

Conclusions

An increase in serum proinflammatory (TNFα, IFNγ), and anti-inflammatory (IL-10) but a fall in IL-4 levels was observed. There was an increase in CD4:CD8 ratio after treatment. Berseem leaves in malnutrition showed comparatively better pro-inflammatory response and reduction in IL-4. Berseem (Leaf) protein offers a good and cheap source of protein with immunomodulating properties to control malnutrition in developing countries.

Further Studies

A CD8 cell rise in number are capable of inducing the death of infected somatic or tumor cells; they kill cells that are infected with viruses. Thus Berseem can be used as a medicine source for treating Cancer and Aids like diseases.

References

Abo-Shousha, S.A., Hussein, M.Z., Rashwan, I.A., Salama, M. *Production of proinflammatory Cytokines: Granulocytemacrophage Colony Stimulating Factor, interleukin-8 and interleukin-6 by Peripheral Blood Mononuclear Cells of Protein Energy Malnourished Children.* Egypt J Immunol 2005; 12: 125-31.

Alvarez-Olmos, M.I., Oberhelman, R.A. *Probiotic Agents and Infectious Diseases: A Modern Perspective on a Traditional Therapy.* Clin Infect Dis 2001; 32: 1567-76. Bistrian BR. Acute Phase Proteins and the Systemic Inflammatory Response. Crit Care Med 1999; 27: 452-3.

Aukrust, P., Müller, F., Ueland, T. Svardal, A.M., Berge, R.K., Froland, S.S. *Decreased Vitamin A levels in Common Variable Immunodeficiency: Vitamin A Supplementation in Vivo Enhances Immunoglobulin Production and Downregulates Inflammatory Responses.* Eur J Clin Invest 2000; 30: 252-9.

Bhaskaram, P., Sivakumar, B.. *Interleukin-1 in Malnutrition.* Arch Dis Child 1986; 61: 182-5.

Biller, H., Bade, B., Matthys, H., Luttmann, W., Virchow, J.C. *Interferon-γ Secretion of Peripheral Blood CD8+ T Lymphocytes in Patients with Bronchial Asthma: in Vitro Stimulus determines Cytokine Production.* Clin Exp Immunol 2001; 126 : 199-205.

Chandra, R.K. *Numerical and Functional Deficiency in T helper Cells in Protein Energy Malnutrition.* Clin Exp Immunol 1983; 51: 126-32.

Cinatl, J., Morgenstern, B., Bauer, G., Chandra, P., Rabenau, H., Doerr, H.W., Glycyrrhizin, *An Active Component of Liquorice Roots, and Replication of SARS-associated Coronavirus.* Lancet 2003; 361 : 2045-6.

Dewan, P., Kaur, I., Chattopadhya, D., Faridi, M.M., Agarwal, K.N. *A Pilot Study on the Effects of Curd (Dahi) & Leaf Protein Concentrate in Children with Protein Energy Malnutrition* (PEM). Indian J Med Res 2007; 126: 199-203.

Doherty, J.F., Golden, M.H., Remick, D.G., Griffin, G.E. *Production of Interleukin-6 and Tumour Necrosis Factor-Alpha in Vitro is reduced in Whole Blood of Severely Malnourished Children.* Clin Sci (Lond.) 1994; 86: 347-51.

Dûlger, H., Arik, M., Sekeroðlu, M.R., Tarakçioglu, M., Noyan, T., Cesur, Y., et al. *Pro-inflammatory Cytokines in Turkish Children with Protein-Energy Malnutrition.* Mediators Inflamm 2002; 11: 363-5.

Giovambattista, A., Spinedi, E., Sanjurjo, A., Chisari, A., Rodrigo, M., Pèrez, N. *Circulating and Mitogen-induced Tumor Necrosis Factor (TNF) Production in Malnourished Children.* Medicina (B Aires) 2000; 60: 339-42.

Jones, P.J. *Clinical Nutrition: 7. Functional Foods- More than Just Nutrition.* CMAJ 2002; 166: 1555-63.

Lopez-Varela, S., González-Gross, M., Marcos, A. *Functional Foods and the Immune System: A Review.* Eur J Clin Nutr 2002; 56 (Suppl 3): S29-33.

Malavé, I., Vethencourt, M.A., Chaæon, R., Quiñones, D., Rebrij, C., Bolívar, G. *Production of Interleukin-6 in Cultures of Peripheral Blood Mononuclear Cells from Children with Primary Proteincalorie Malnutrition and from Eutrophic Controls.* Ann Nutr Metab 1998; 42: 266-73.

Mediratta, P.K., Dewan, V., Bhattacharya, S.K., Gupta, V.S., Maiti, P.C., Sen, P. *Effect of Ocimum Sanctum Linn. on Humoral Immune Responses.* Indian J Med Res 1988; 87 : 384-8.

Rodríguez, L., González, C., Flores, L., Jiménez-Zamudio, L., Graniel, J., Ortiz, R. *Assessment by Flow Cytometry of Cytokine Production in Malnourished Children.* Clin Diagn Lab Immunol 2005; 12 : 502-7.

Solis, B., Nova, E., Goìmez, S., Samartiìn, S., Mouane, N., Lemtouni, A., et al. *The Effect of Fermented Milk on Interferon Production in Malnourished Children and in Anorexia Nervosa Patients undergoing Nutritional Care.* Eur J Clin Nutr 2002; 56 (Suppl 4): S27-33.

OO

Food Heritage of India

Aparna Raj and D.K.Bhatt

Abstract

This paper analyses the transformation and redefinition of local identity in rural India from the perspective of heritage - more precisely food and gastronomy - and local rural tourism. As an identity marker of a geographic area and/or as a means of promoting farm products, gastronomy meets the specific needs of consumers, local producers and other actors in rural tourism. The paper considers the meaning of food from a theoretical perspective. The current interest in traditional food and cuisine is part of a general desire for authentic experiences. At the regional level, the dynamics of building up heritage consist in actualizing, adapting, and re-interpreting elements from the past, thus combining conservation and innovation. Local development can be seen as a process of territorial and heritage construction. Culinary heritage is a social construction and an important resource for local action. Varied, healthful, enticingly aromatic and abundantly flavorful, Indian food is fast becoming the world's most devoured ethnic cuisine. The unrivaled collection of Indian recipes span the spectrum of regional specialties The present paper elaborates the gastronomic journey which can be taken by a tourist on his visit to India.

Introduction

Food and beverage tourism is increasingly being recognized as an important part of the cultural tourism market. This can include gastronomic tourism, culinary tourism, and cuisine tourism. At the same time, special events and festivals have become one of the fastest growing types of tourism attractions and have attracted attention from tourism marketing professionals. The interest in such experiences also has fueled the development of food festivals that rural communities use to promote local products and rural food heritage and differentiate themselves from urban community festivals.

Food Tourism is closely linked with Agro Tourism. Agro Tourism is when a native person or local of the area offers tours to their Agriculture project to allow a person to view them growing, harvesting, and processing locally grown foods, such as coconuts, pineapple, sugar cane, corn, or any produce the person would not encounter in their home country. Often the farmers would provide a home-stay opportunity and education.

The essential benefit is for the local people, not tour operators or hotels. People need to earn money to live and support their family, but often this is at the cost of the environment and the planet. Agro Tourism and Food Tourism project the heritage of a country. A Country like India has an extremely varied and authentic variety in Food and Drinks and it can be projected as a place which has a rich food heritage.

Many who attempt to define food and culinary tourism immediately think of wineries and fine restaurants. These are two components of the niche, but by no means a definitive list. Food tourism can occur at a farmers' market, or even in the home of a friend or relative. Travellers do not often choose their holiday destination based on the food tourism experiences they anticipate to encounter, but still end up remembering their holiday to a certain extent on the quality of the food they experienced at the destination. This creates a conundrum in that many tourists choose holiday destinations based on one perceived aspect (e.g. beaches, accommodation etc), but their actual satisfaction will be based on aspects (i.e., food) they did not consider in their original holiday choice.

Food tourism can essentially be viewed as a subset of cultural tourism, with the local cuisine being a product of the local culture and the natural environment. Therefore, regions that possess unique dishes and food products as a result of their culture and environment may be transformed into food tourism destinations with minimum marketing and product development.

The finest of India's cuisines is as rich and diverse as it's civilization. It is an art form that has been passed on through generations purely by word of mouth. The range assumes astonishing proportions when one takes into account regional variations. Very often the taste, colour, texture and appearance of the same delicacy changes from state to state.

The hospitality of the Indians is legendary. In Sanskrit Literature the three famous words 'Atithi Devo Bhava' or 'the guest is truly your god' are a dictum of hospitality in India. Indians believe that they are honoured if they share their mealtimes with guests. Even the poorest look forward to guests and are willing to share this meager food with guest. And of particular importance is the Indian woman's pride that she will not let a guest go away unfed or unhappy from her home. Indians are known for their incredible ability to serve food to their guests invited or uninvited.

Food customarily forms the crowning part of most festivities and celebrations. Whatever the occasion Indians eat with great gusto and are adept at finding reasons to feast and make merry. At traditional and festive meals, the thali (plate) or banana leaf is decorated with rangoli (a design drawn with white and colored powders around the edges).

Indian cooking is one of the great cuisines of the world. Like the country itself, however, it varies greatly from region to region, and one can discover a great deal more to savor than the ubiquitous *kormas* and *tikka* masalas (known to the naive simply as "curry") with which most Westerners are familiar. Not only does each Indian community and ethnic and regional group have a distinct cuisine, but there is a great deal of fusion within the country — subtle variations and combinations you're only likely to pick up once you are familiar with the basics. A good way to sample a variety of dishes in a particular region is to order a *thali* (multicourse meal), in which an assortment of items is served. Basic staples that tend to be served with every meal throughout the subcontinent are rice, *dal* (lentils), and/or some form of *roti* (bread).

Literature Review

Food has been associated with travel from prehistoric times (Boniface, 2003). Changes in the ways in which food was obtained in ancient times transformed the status of food from a basic survival necessity to a commodity associated with wealth, celebration, rituals and leisure (Tannahill, 2002). At the same time, the exchange of food stuffs over long distances and the transfer of food-related ideas and knowledge across cultures assisted in the construction of cultural identities and social hierarchies (Pilcher, 2006). These historical events were interpreted by Boniface (2003) as a "precursor" that laid the foundation for "culture, food and drink, and tourist tourism to feature and interact together"

The relationship between food and tourism has progressed from "traditional hospitality, cuisine and gastronomy" to the development of the innovative concept of "food tourism" (Jones & Jenkins, 2002, p. 115), also referred to as "culinary", "gastronomic" or "gourmet" tourism (Okumus et al., 2007, p. 19). Today, food is an integral part of the overall tourism experience, in addition to it being a prime motivation for travel (Hall, Mitchell, & Sharples, 2003). Hall and Mitchell (2005) with reference to Hall and Mitchell (2001) defined food tourism as, "visitation to primary and secondary food producers, food festivals, restaurants and specific locations ... it is the desire to experience a particular type of food or the produce of a specific region".

In the selection process of a destination, food presents tourists with an experience that includes excitement, cultural exploration and inspiration (Scarpato & Daniele, 2003; Sharples, 2003). In fact, it is argued that food is the most important attribute after climate, accommodation and scenery, in choosing a destination, (Hu & Ritchie, 1993). Consumption of food is believed to provide unforgettable tourist experiences (Law & Au, 2000), the memories of which are recalled long after the holidays are over (Ravenscroft & Westering, 2002).

Indian Food Heritage

Indian food is one of the great cuisines of the world. Like the country itself, however, it varies greatly from region to region. Not only does each Indian community and ethnic and regional group have a distinct cuisine, but there is a great deal of fusion within the country — subtle variations and combinations you're only likely to pick up once you are familiar with the basics. A good way to sample a variety of dishes in a particular region is to order a *thali* (multicourse meal), in which an assortment of items is served. Basic staples that tend to be served with every meal throughout the subcontinent are rice, *dal* (lentils), and/or some form of *roti* (bread). The following is a brief summary of regional variations and general dining tips.

Southern States—Food from the coastal areas of India almost always contains a generous quantity of coconut — besides using it in cooking, most Maharashtrian homes offer grated coconut as a garnish to every dish. Rice also dominates the food of southern India, as do their "breads," which are more like pancakes and made of a rice (and/or dal) batter — these *appams, idiapams*, and *dosas* are found throughout the south. *Dosas* are in fact a South Indian "breakfast" favorite (consumed anytime), as are *idlis* and *vadas*, all of which have become part of mainstream cooking in many parts of India. *Idli* is a steamed rice and lentil dumpling, *dosa* a pancake (similar batter), and *vada* a deep-fried doughnut-shaped snack. All should be eaten fresh and hot with a coconut chutney and *sambar*, which is a specially seasoned *dal* (lentils), also eaten with steamed rice. In Tamil Nadu a large number of people are vegetarian, but in Kerala, Goa, and Mumbai, you must sample the fresh fish! Delicious kebabs and slow-cooked meals are what you'll find in Hyderabadi cuisine; inspired by the courts of the *Nawabs* (nobles), it's similar to Mughlai cooking, but stronger in flavor.

Northern States — India's great meat-eating tradition comes from the Mughals and Kashmiris, whose *rogan josh* and creamy *korma* dishes, along with kebabs and *biryanis*, have become the backbone of Indian restaurants overseas. The most popular tradition—tandoor (clay oven) cooking—is part of India's Mughal gastronomic heritage. Tandoor dishes are effectively "barbecued" vegetables, *paneer* (Indian cheese), or meat that has been marinated and tenderized in spiced yogurt, cooked over coals, and then either served "dry" as a kebab or in a rich spiced gravy like the *korma*. Recently revived is the tradition of *dum pukht*, enjoyed by the erstwhile *Nawabs* of Awadh in Lucknow and the surrounding area. All the ingredients are sealed and slow-cooked in a pot, around which coals are placed. Nothing escapes the sealed pot, preserving the flavors.

Northwest (Punjab) Specials—Besides trying the various tandoor dishes, you should order *parathas:* A Punjabi specialty, this thick version of the traditional chapati is stuffed with potatoes, cabbage, radish, or a variety of other fillings. Be aware that many North Indians love their ghee (clarified butter); sensitive stomachs (or those watching their weight) should simply specify that they would prefer their *paratha* without ghee. A general note of caution when dining in North India: If the menu specifies a choice between oil and ghee as a cooking method, you should probably specify the former. And keep in mind that if you exclusively eat oily, highly pungent, so-called Punjabi fare, you are bound to feel ill, so make sure you vary your meals by dining at South Indian restaurants, which combine a healthy balance of carbohydrate and protein (rice and dal); in northern states you will find *rotis* (breads) combined with *rajma* (kidney beans), *puris* (bread) with *chole* (chickpeas), and so on.

Eastern States—Freshwater fish (such as *hilsa, bekti*, and *rohu*) take pride of place at the Bengali table, which incidentally considers itself to be the apotheosis of Indian cooking. In Bengal, mustard oil (which has its own powerful flavor) is the preferred cooking oil. Sweets are another Bengali gift to the world; these are made from milk that has been converted to *paneer* (Indian cheese) and that has names like *rosogolla* (or *rasgulla*) and *sandesh.*

Spices—Literally hundreds of spices (masalas) and spice combinations form the culinary backdrop to India, but a few are used so often that they are considered indispensable. Turmeric *(haldi)*—in its common form a yellow powder with a slightly bitter flavor — is the foremost, not least for its antiseptic properties. Mustard seeds are also very important, particularly in the South. Cumin seeds and coriander seeds and their powders are widely used in different forms—whether you powder, roast, or fry a spice, and how you do so, makes a big difference in determining the flavors of a dish. Chili powder is another common ingredient, available in umpteen different varieties and potencies. Then there are the vital "sweet" spices—cardamom *(elaichi)*, clove *(lavang)*, cinnamon *(dalchini)*—which, along with black pepper *(kali miri)*, make up the key ingredients of the spice combination known as *garam masala.* Though tolerance to spicy food is extremely subjective, let your preference be known by asking whether the item is spicy-hot *(tikha hai?)* and indicating no-chili, medium-spicy, and so on. "Curry powder" as it is merchandised in the West is rarely found or used in India. "Curry" more or less defines the complex and very diverse combination of spices freshly ground together, often to create a spicy saucelike liquid that comes in varying degrees of pungency and varies in texture and consistency, from thin and smooth to thick and grainy, ideally accompanied by rice or breads.

Staples & Accompaniments — All over the country, Indian food is served with either the staple of rice or bread, or both— the most popular being unleavened (pan-roasted) breads (called *rotis*); tandoor-baked breads; deep-fried breads (*puris* and *bhaturas*) or pancake-style ones. Chapatis, thin whole-wheat breads roasted in a flat iron pan *(tava)*, are the most common bread eaten in Indian homes, though these are not as widely available as restaurant breads. The thicker version of chapatis are called *parathas*, which can be stuffed with an assortment of vegetables or even ground meat. Tandoor-roasted breads are made with a more refined flour and include *naans*, tandoori *rotis*, and the super-thin *roomali* (handkerchief) *rotis*. Tandoor breads turn a little leathery when cold and are best eaten fresh.

Dal, made of lentils (any of a huge variety) and seasoned with mustard, cumin, chilies, and/or other spices, is another Indian staple eaten throughout the country. *Khichdi*, a mixture of rice, lentils, and spices, is a great meal by itself and considered comfort food. In some parts it's served with *kadhi* — a savory sour yogurt-based stew to which chickpea flour dumplings may be added. You'll usually be served accompaniments in the form of onion and lime, chutneys, pickles, relishes, and a variety of yogurt-based salads called *raita. Papads* (roasted or fried lentil flour discs) are another favorite food accompaniment that arrive with your meal in a variety of shapes, sizes, and flavors.

Meat — A large number of Indians are vegetarian for religious reasons, with entire towns serving only vegetarian meals, but these are so delicious that meat lovers are unlikely to feel put out. Elsewhere, meat lovers should probably (unless you're dining in a top-end big-city restaurant) opt for the chicken and fish dishes — not only are these usually very tender and succulent, but the "mutton" or "lamb" promised on the menu is more often than not goat, while "beef" (seldom on the menu — beef is taboo for most Hindus, and the ban on cow slaughter continues to be a raging national debate) is usually water buffalo. Again, there are regional differences, like in "Portuguese" Goa, where pork is common.

Sweets — Indians love sweets (called *mithais, mishtaan*, or "sweet meats'"), and they love them very sweet. In fact, Western palates often find Indian sweets *too* sweet; if this is the case, sample the dry-fruit-based sweets. Any occasion for celebration necessitates a round of sweets as a symbol of spreading sweetness (happiness). Every region of the country has a variety of specialty sweets made from an array of ingredients, but they are largely milk-based. This includes *pedas* and *laddus* (soft, circular), *barfis* (brownielike), *halwas* (sticky or wet), *kheer* (rice pudding-like), and so on. Whatever you do, don't miss the Indian *kulfi*, a creamy, rich ice cream flavored with saffron, nuts, or seasonal fruit.

Fruits — If the spiciness of the meals unsettles your stomach, try living on fruit for a day. You'll get a whole range of delicious tropical varieties (with any luck, in a basket in your hotel room) ranging from guavas and jackfruit to lychees and the most coveted fruit of them all, the mango. More than 200 varieties of mangoes are grown in India, but the most popular ones (Alphonso or *aphoos*) come from Maharashtra (Mar-June); try to taste one when you're in Mumbai.

Beverages — *Chai* (tea) is India's national drink. Normally served in small quantities, it is hot, made with milk usually flavored with ginger and/or cardamom, and rather sweet unless you request otherwise. Instant coffee is widely available (and may be mixed in your five-star hotel's "filter coffee" pot), but in South India you'll get excellent fresh brews. Another drink worth trying is *lassi*, liquefied sweetened yogurt. ***Note:*** The yogurt is sometimes thinned with water, so you're only safe consuming lassi in places where they can assure you no water was added at all, or where they will make it with bottled water (that you purchase separately). Lassi's close companion is *chaas*, a savory version that is very thin and served with Gujarati/Rajasthani meals. With southern food, it is served with a flavorful assortment of herbs and spices. In general, you should avoid ice in any beverage unless you are satisfied that it is made from boiled water.

References

Boniface, P. (2003). *Tasting Tourism: Travelling for Food and Drink.* Hampshire: Ashgate Publishing Limited.Tannahill, 2002

Pilcher, J.M. (2006). *Food in World History.* Oxon: Routledge.

Jones, A., & Jenkins, I. (2002). '*A taste of Wales - Blas Ar Gymru*': Institutional malaise in promoting Welsh food tourism products. In A.M. Hjalagar & G. Richards (Eds.), Tourism and Gastronomy (pp. 114-131). Oxon, UK: Routledge.

Okumus, B., Okumus, F., & McKercher, B. (2007). Incorporating local and international cuisines in the marketing of tourism destinations: The cases of Hong Kong and Turkey. Tourism Management, 28, 253-261.

Hall, C.M., & Sharples, L. (2003). *Consuming Places: The Role of Food, Wine and Tourism in Regional Development.* In C.M. Hall, L. Sharples, R. Mitchell, N. Macionis & B. Cambourne (Eds.), Food Tourism Around the World: Development, Management and Markets (pp. 25-59). Oxford, UK: Butterworth- Heinemann.

Hall, C.M., & Mitchell, R. (2005). *Gastronomic Tourism: Comparing Food and Wine Tourism Experiences.* In M. Novelli (Ed.), Niche tourism: contemporary issues, trends and cases. Oxford, UK: Butterworth-Heinemann.

Scarpato, R., & Daniele, R. (2003). *New Global Cuisine: Tourism, Authenticity and Sense of Place in Postmodern Gastronomy.* In C.M. Hall, L. Sharples, R. Mitchell, N. Macionis & B. Cambourne (Eds.), Food tourism around the world: Development, management and markets. Oxford: Butterworth-Heineman.

Law, R., & Au, N. (2000). *Relationship Modelling in Tourism Shopping: A Decision Rules Induction Model.* Tourism Management, 21(3), 241-249.Hu, Y., & Ritchie, J.R.B. (1993). Measuring destination attractiveness: A contextual approach. Journal of Travel Research, 32(2), 25-32.

Sharples, L. (2003). Food tourism in the peak district national park, England. In Food tourism around the world: Development, manage- ment and markets. Oxford, UK: Butterworth-Heinemann.

Ravenscroft, N., & Westering, J.V. (2002). *Gastronomy and Intellectual Property.* In A.-M. Hjalager & G. Richards (Eds.), Tourism and gastronomy (pp. 153-165). London: Routledge.

OO

23

Exploring Challenges and Opportunities for Sustainable Development of Agro-enterprise in India: A Value Chain Approach

Mohd. Shamim Ansari

Abstract

The problem in agriculture is lack of low value addition. Much of the product is sold raw, value addition reaches in cities often reaching back the village at a higher price. This can be avoided by setting value addition facilities at village and marketing value added products to cities at a higher price. Value addition as a rural enterprise has a potential to generate more local jobs, better income and reduce rural migration.

Introduction

"All people are entrepreneurs, but many don't have the opportunity to find that out"

—Muhammad Yunus

Low level of income for farmers is one of the basic problems of India agriculture. This could be attributed to farmers' inability in grabbing opportunities of value addition as a primary supplier in the food sector. They wrestle with several challenges viz. uncertain weather, depleted natural resources, unpredictable market situation, insensitive policies, small land holding, poor access to technical know-how and credit. Widening gap between prices paid by consumers for value added products and farmer's realisation is one of key reasons for poor shape of rural enterprise in India though it has tremendous potential.

Farmers' earning can be raised not just by increasing productivity but also through efficient and effective value addition. Realizing the value chain potential can help farmers to market their products, sharpen their entrepreneurial skills and reduce migration. Indian Farmers today need exposure to new ideas and possibilities, understand and operationalise those ideas that would help in forming business and take control of the value chain locally. The value chain approach covers every step from the farmer to the final consumer where each step would add value. This would help farmers to fetch better price for their produce. However, value addition is not new for Indian farmers but has been only at individual level. In order to achieve economies of scale community base value addition is required which could be possible through forming Self Help Groups.

A value chain approach can be one of the way for a firm to develops competitive advantage and creates shareholders value. Value chain allow for examining the consequences of empowering on group (farmers) and identifying how to link them to consumers of domestic and foreign market.

In this back drop this paper aim at evaluating Strengths, Weaknesses, Threats and exploring of opportunities for development Agro-enterprise in India. An attempt has been made to emphasise how value chain approach can be fruitful in transforming farmers as agripreneurs.

Concept of Value Addition

Adding value is the process of changing or transforming a product from its original state to a more valuable state. A broad definition of value addition is to economically add value to a product by changing its current place, time and from one set of characteristics to other characteristics that are more preferred in the marketplace. The value of a changed product is added value, such as processing wheat into flour and then into bakery products such as bread, pastry, cake etc. desired by consumers. The Value to customers basically emphasises upon ratio of perceived benefits and costs

It is important to identify the value-added activities that will support the necessary investment in research, processing and marketing. The application of biotechnology, the engineering of food from raw products to the consumers and the restructuring of the distribution system to and from the producer all provide opportunities for adding value. The value chain look likes following.

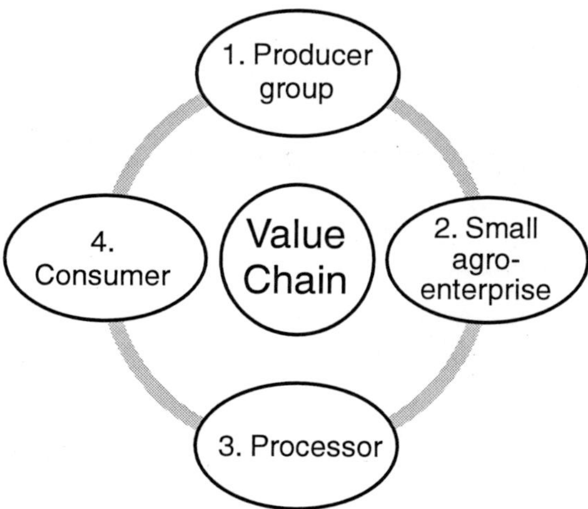

Approaches to Adding Value

In general, the problem is to evaluate what, where, how and who can efficiently perform the marketing functions. The approach to adding value can be classified into (i) Innovation and (ii) Coordination.

Innovation	*Coordination*
Innovation focuses on improving existing processes, procedures, products and services or creating new ones. Innovation also can come from research about alternative crops that can be grown successfully by producers to replace traditional crops. Value-added producers are able to economically profit by growing these alternative crops instead of traditional crops. Some alternative crops that show promise include industrial hemp for its floriculture for oil, horticulture for fruit juice etc. Another type of innovation is industrial innovation, where the processing of traditional crops takes place into non-food end uses such producing ethanol from corn, biodiesel from soybeans and particleboard from straw.	Coordination focuses on arrangements among those that produce and market farm products. This can be both vertical and horizontal. Horizontal coordination involves pooling or consolidation among individuals or companies from the same level of the food chain. Vertical coordination includes contracting, strategic alliances, licensing agreements and single ownership of multiple market stages in different levels of the food chain. It is necessary to link production processes and product characteristics to the preferences of consumers and processors.

Significance of Value Addition

Whether you capture value or create value, the bottom line is that you get paid for providing value. If your business venture does not provide value to the system, there is no reason to expect a return. So the process of creating a successful business involves the search for providing value. Providing value can be in the form of marketing a unique product, filling a market niche, simplifying the supply chain, providing a service, lowering costs, and many other ways. The more value you provide, the more return you can extract from the marketplace. Value added approach enables product differentiation and innovation which in term helps in creating competitive advantage for capturing larger market share.

Problems of Value Addition Faced by Indian Farmer

1. Farmers in India generally do not go for grading of products and sell mixed different qualities. Thus, they lower price.
2. Farmers are forced to sell the perishable produce straight after harvest at a low price because they don't have cheap and effective storage facilities.
3. Farmer are generally in debt to local money lender and traders and so had to sell their produce immediately to pay back their loans.

4. Use of Different types of seeds, planting times and growing techniques lead to harvest at different times, thereby increasing cost for traders and reducing quality.

Some of the other basis issues that create hurdle for effective value addition for agro-enterprise industries in India could be summed up as follows.

Low level of value addition	Bottle necks in cold storage and during transit Seasonality – capacity utilization issues Non Efficient storage/warehousing, processing & marketing techniques Non adoption of efficient technology
Infrastructure and others	Shortage of power High electricity tariff Low area under Irrigation High capital cost – High insurance premium in risk coverage Farm connectivity by road yet to take off
Quality and consistency	Inputs delivery not in time Innumerable varieties Poor procurement Lack of cheap and timely credit Requirement of higher working capital
Weak and ineffective supply chain	Non professional management Low revenue rentals Non Efficient and competitive retailing High wastage Too much Dependence on intermediaries

The **Figure 23.1** and **Figure 23.2** presented below illustrate the comparative difference between supply chain in India and developed countries.

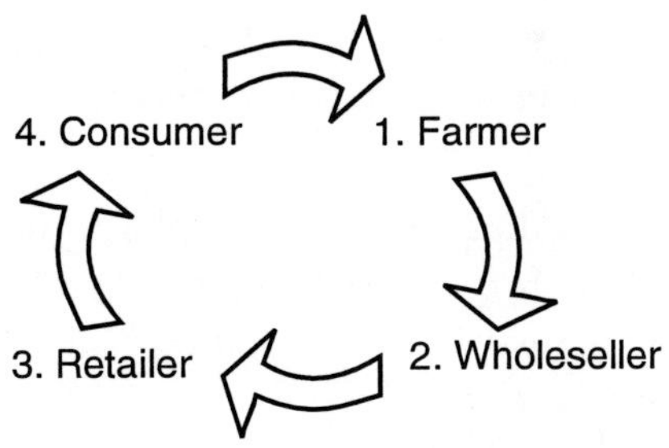

Figure 23.1: *Supply Chain Developed Countries*

Note: Shorter supply chain result into low waste and high margin for farmers.

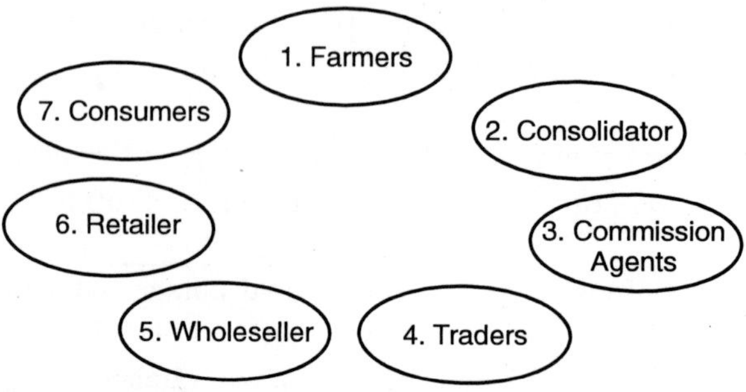

Figure 22.2: *Supply Chain in India*

Note: Longer supply chain result into high wastage and low margin for farmers

Routes for Value Creation

I. *Innovation at the Consumer end of the Chain:* This requires Product and Purchase experience improvements through processing, packaging & retail format innovations, delivering variety, quality, food safety, shopping convenience, better shelf life, ready to eat etc. It would help in accelerating demand and fetch price premiums.

II. *Efficiency along the whole chain:* This requires Lower industry cost structures through integration, coordination, disintermediation, infrastructure investments etc. It would help in pricing flexibility and traceability.

Opportunities

"Produce-and-then-sell" mentality of the commodity business is being replaced by the strategy of first determining what attributes consumers want in their food products and then creating or manufacturing products with those attributes. With the continuous shifting to a global economy, the international market for value-added products is growing. Market forces have led to greater opportunities for product differentiation and add value to raw commodities because of:

1. Increased consumer demands regarding health, nutrition and convenience;
2. Efforts by food processors to improve their productivity; and
3. Technological advances that enable producers to produce what consumers and processors desire.

Producers involved with adding value must thinks beyond commodity producers. Value chain approach can helps in absorbing risk involved in agriculture. Value chain approach enable selling value added food product rather than simply selling raw commodities. This could range from:

1. Adding value to cattle, fish and poultry products
2. Marketing crops like organically grown grains, potatoes, carrots, beans, tomatoes and corn etc.

Producers have a challenge to be responsive to consumer demands by producing what is desired. Attentiveness to consumer demands in quality, variety and packaging are important, because demographic trends show growth in the convenience-oriented, health-conscious and environmentally concerned sectors where price is not as important as quality.

Conclusion

Value chain approach can help in achieving competitive advantages, economies of scale and economies of scope. It may come from being able to sell at the lowest price due to scale economies, having a monopoly, or being among the first to produce or market in a new way. However, farmer are not able to find advantage due to following challenges which must be addressed on priority.

Challenges	*Ways out*
Majority of the farmers are small & resource-poor	Many studies have revealed that local enterprise work best if they form themselves, and develop there own rules. They need to be inducted in to the new structure responsibly. Formation of SHGs can help in development of agro-enterprise industries.
Physical infrastructure & new market institutions are still evolving	This needs to be appropriately factored in. Corruption, unfair taxation or bureaucracy can create a hurdle in flourishing small business. An association of farmers and several enterprise can create a favaourable small business environment.
Understanding of most of the agro business players is insufficient leading to disputes	An agreement should be made between producers and enterprise signed by a third party such as local body for economical and easy arbitration.
Clash of local interest and start up capital	To ensure that agro-enterprise stays locally owned, outside companies should be restricted from buying shares or having voting rights. Donor co-funding during the business planning stage is also an effective way to enable small agro-enterprise to invest and maintain their independence.
• Control of markets by few traders and agents who have large storage capacity • Crashing of prices during peak season discourage the farmers • Larger intermediaries leads to shrinkage of margin available to farmers. • Multiple handling deteriorates the quality of product	In order to address these challenges a mechanism need to be developed for providing organisational support to new enterprises for several years. Government must be instrumental to address these issues. Giving tax free start up periods on condition that profits are reinvested in the enterprise and can encourage the growth of new enterprise. Supply chain model have to shortened by eliminating the middlemen.

References

Akridge, J., D. Downey, M. Boehlje, K. Hariing, F. Barnard, and T. Baker. 1997. "*Agricultural Input Industries*." Food System 21 Gearing Up for the New Millennium, Chapter 15. Purdue University Cooperative Extension Service, West Lafayette, Indiana.

Barkema, A., and M. Drabenstott. 1996. "*Consolidation and Change in Heartland Agriculture*." Economic Forces Shaping the Rural Heartland. Federal Bank of Kansas City, Missouri.

Barkema, A., and M. Drabenstott. 1995. "*The Many Paths of Vertical Coordination: Structural Implications for the US Food System*." Agribusiness, 11(5).

Dobbs, T., M. Leddy, and J. Smolik, 1988. "Factors influencing the economic potential for alternative farming systems: case analysis in South Dakota." *American Journal of Alternative Agriculture* 3(1): 26-34.

Connell, J.G. and O. Pathammavong, 2007. "*Starting and Agro-enterprise development Process*." Field Facilitators Guide, CIAT Asia.

Rodener, Daniel. 2007. "*Donor Interventions in Value Chain Development*." Swiss Agency for Development and co-operation, Berne, Switzerland.

24

Growth, Potential and Innovation in Food Processing Industry in India

Dr. C.B. Singh

Abstract

India is the world's second largest producer of food next to China, and has the potential of being the biggest in the world. Food processing is a key industrial sector for India; it accounts for a gross output of more than US $ 69.4 billion, out of which value-added food products comprise US $ 22.2 billion. The sector employs about 13 million people directly and 35 million people indirectly. Size of the semi-processed and ready to eat packaged food industry is over US $ 1 billion, and it is growing at over 20 per cent a year. An inefficient agric supply-chain, inadequate post-harvest infrastructural facilities, lack of adequate market access, insufficient and high rate of credit make agriculture a risky proposition are major hurdles in the development of food industry. This study has been done on the basis of secondary data. Secondary data was collected from journals, books, magazines and Internet as well as government agencies.

This paper is an attempt to identify an opportunity for large investments (domestic and FDI) in food and food processing industry, emergence of organized Retail Sector in India. Identify the major factors fuelling this change. Initiatives undertaken by the Government of India to drive consumption, efforts made to attract domestic and international players in this sector. The industry now required more advanced infrastructure facilities for frozen fruits and vegetable products and seafood and meat products. Today advanced farmers are earning more money through applying Information Technology, contact farming, organized sale and modern technology.

Keywords: *Organized retail, infrastructure, and disposable income food processing industries and supply chain.*

Introduction

India is the world's second largest producer of food next to China, and has the potential of being the biggest in the world. Food processing is a key industrial sector for India; it accounts for a gross output of more than US $ 69.4 billion, out of which value-added food products comprise US $ 22.2 billion. The sector employs about 13 million people directly and 35 million people indirectly. Size of the semi-processed and ready to eat packaged food industry is over US $ 1 billion, and it is growing at over 20 per cent a year. The food industry in India comprises food production and the food processing industry. The food processing industry is one of the largest in India – it is ranked fifth in terms of production, consumption, export and expected growth.

- Agriculture sector is vital for any nation and India is the principal source of livelihood for more than 58 per cent of the population. Agriculture sector has touched a growth rate of 4.4 per cent in the second quarter of 2010-11 thereby achieving an overall growth rate of 3.8 per cent during the first half of 2010-11. The food processing industry in India is at an early growth stage, with low penetration levels and high potential.
- The size of the food processing industry in India has increased from US$ 57 billion in 2004 to US$ 75 billion in 2007. During this period, the number of registered operating units increased from 24,000 to 25,725 units. India's processed food exports constituted 1.5 per cent of the global food trade in 2008–09.
- Area under food crops has increased from 122.78 million hectare (ha) in 2001-02 to 125.73 million ha in 2010-11 (4th advance estimate). Production of food grains has increased from 212.85 million tonne (MT) in 2001-02 to 241.56 MT during 2010-11(4th advance estimates). The food grain production target for the year 2011-12 has been fixed at 245 MT, which is likely to be achieved on back of favourable weather conditions.

The Indian Agriculture and food industry can be categorized into five broad segments, namely:

- Fresh fruits and vegetables
- Floriculture, comprising fruit and vegetable seeds and flowers
- Processed fruits and vegetables and other processed foods
- Animal products, including meat, poultry, dairy and honey
- Cereals such as rice and wheat

The key focus areas of Agro/Food are:

- Emergence of Organized Retail Sector in India and Food and Food Processing industries
- Processing industry and value addition
- Building backward/forward linkages between farmers and producers.
- More advanced infrastructure facilities for frozen fruits and vegetable products and
- Seafood and meat products Development of "food processing clusters"
- Establishment of Certification and testing centers in Tamil Nadu.
- Skill development for food processing personnel

- Technology interventions and R&D
- Identifying Export Opportunities –marketing structures.
- Adopting "best practices" in food processing from other countries

The food processing industry is very important for an agriculture-based economy like India because it helps in the commercialization of farming and increases the income of farmers. It also generates employment opportunities and assists in the creation of markets for export of agriculture-based products. In the beginning, the food processing industry was limited to procedures of food preservation and packaging that involved drying, salting, and pickling. However, in the last few years, with advancement in technology, the scope of the sector has grown tremendously. The industry now also includes ready-to-eat food items, frozen fruit and vegetable products, and seafood and meat products. The storage, processing, preservation, and transportation of various food items have given rise to many irradiation facilities (Food irradiation is the process of exposing food to ionizing radiation to destroy microorganisms, bacteria, viruses, or insects that might be present in the food.), cold storage facilities, and packaging centers.

Food Processing Industries and its Advantages to India

(1) India's tropical climate favours the cultivation of several exotic food and flower crops. The peninsular coastline of the country drives growth of the marine industry.

(2) India is the largest producer of several fruits, such as banana, mango and papaya, and the second-largest producer of vegetables such as brinjal, cabbage and onion. Further, India is also one of the largest producers of rice.

(3) India has the largest livestock population in the world with 98.7 million buffaloes and 176 million cows (2008).

(4) The sector employs about 13 million people directly and 35 million people indirectly. The presence of several agricultural institutes, such as the Indian Agricultural Research Institute (IARI), serves as a perfect platform for research and innovation.

(5) The Middle East and Southeast Asia are major export destinations for Indian agricultural (agri) commodities and milk.

(6) The establishment of 60 fully equipped Agri Export Zones (AEZs), in addition to food parks, act as an incentive for attracting foreign investment.

An inefficient agri supply-chain, inadequate post-harvest infrastructural facilities, lack of adequate market access, insufficient and high rate of credit make agriculture a risky proposition are major hurdles in the development of food industry.

Objective of Study

(1) To know the market size and growth drivers of Indian food Industry.

(2) Opportunity for large investments (domestic and FDI) in food and food processing industry

(3) Opportunities for Emerging Organized Retail Sector In India

(4) Identifying Export Opportunities –marketing structures.

Research Methodology

This study work is completely based on secondary data. The secondary data and relevant material were collected from various official and non-official sources: various sources are as following:

- Data analyzed for WTO agreement and its impact on global food security taken World Bank (2007) Agricultural Development Report 2008. USA, Washington, DC.
- Flavors of Incredible India, Ernst & Young, 2009
- Corporate Catalyst India (CCI) Survey Report, Press Releases,
- KPMG Report on Food Processing and Agri Business,
- Agriculture and Processed Food Products Export Development Authority (APEDA) articles,
- Ministry of Food Processing Industries articles,
- RNCOS Research Report, Department of Industrial Policy and Promotion (DIPP), Media Reports
- Cygnus Research, Oct 2008 www.ibef.org

Result and Discussion

1. *Market Size of Indian Food Industry*

A. Food processing industry

Food processing industry is of enormous significance for India's development as it has linked economy, industry and agriculture in India, efficiently and effectively. The three pillars being together have synergized the development process and promoted the growth of the nation to a great extent.

There are 25, 367 registered food processing units in the country whose total invested capital is Rs 84,094 crore (US$ 17.81 billion), as per a competitiveness report of the National Manufacturing Competitiveness Council. This information was given by Dr Charan Das Mahant, the Minister of State for Food Processing Industries, in a written reply to the Lok Sabha.

The food processing sector is presently growing at an average rate of 13.5 per cent per annum. The Vision Document 2015 envisages increasing the value addition from 20 per cent to 35 per cent by 2015.

Food processing industry is one of the largest industries operating in India and is divided into several segments.

B. Key drivers of growth of Indian retail sector

By 2015, the Indian food industry is estimated to grow by about 40 per cent over the level in 2007. This growth is expected to be driven by two key factors

- **Socio-economic changes:** Across India's population base, in terms of growth in the number of households in the higher income category, increasing youth population and migration from rural to urban areas.
- **Changing and evolving lifestyle trends:** Such as emergence of nuclear families, increasing health aware- ness and growing exposure to international markets
- **Higher disposable income:** Many studies are highlighting the fact that increase in disposable income has helped in boosting retail growth in India. Entry of foreign firms here has increased the salary competitiveness, thus pressurizing Indian firms.
- **Demographic transition:** The demographic change in India is much talked about subject. The median age for India is around 25 years in 2009 (Central Intelligence Agency estimates). The proportion of population in the age group of 25-59 years has been gradually increasing and is likely to increase further to 23% of the total population by 2010.
- **Young age and income earning capability:** Young age and income earning capability is a potential attraction which very few countries offer.

  Young people with income earning capacity are more experimental (more likely to visit a mall), seek con- venience (demand for process food) and less concerned about saving though are value conscious.
- **Growing working women population:** The propensity to spend in the case of working women is higher by 1.3 times as compared by housewives. According to the census report, the population of working women increased to 18 per cent in 2005 as compared to 12 per cent in 1991. Increasing proportion of working women means kitchen getting redefined as well as increasing demand for more processed/semi-processed food, convenience seeking pre-mix, time saving recipes and so on. Also, working women value time, space and privacy. With both the husband and wife working they look for one-stop shopping, speed and efficiency of purchases that lacks in traditional system.
- **Growth in urban population:** If we look at rural retailing the penetration of modern retail is very minute. The shopping pattern changes with urbanization. Share of rural population in total population has decreased from 74 per cent in 1990 to 70 per cent in 2008.The migration toward cities for work and education is a growing phenomenon among rural areas.

By 2026, India's urban population is projected to become 38.2 % of total population.

- **Increasing private final consumption expenditure:** The growth in PFCE was in line with growth of GDP till 2007-08, interestingly though the GDP growth has come down; the PFCE growth rate is constant at 12%. The trend indicates increasing and resilient consumer spending which would favour consumption growth and retail.

C. Exports

Exports of organic food products are expected to grow five-fold by 2015, according to the Agriculture and Processed Food Products Export Development Authority (APEDA).The Government agency expects exports to touch US$ 1.43 billion by 2014-15 against US$ 280 million in 2010-11."There are a total of 2,084 organic projects in India which have been certified by our certification bodies. 191 out of these are exporters of organic produce," as per Shailender Singh, Consultant, Organic Division, APEDA.

TABLE 24.1

Exports of Processed Food (US$ million)

Year	*Exports of processed food (US$ million)*	*Growth %*
2003-04	600.1	--------
2004-05	619.9	3.32
2005-06	1153.3	86.27
2006-07	1390.5	20.57
2007-08	1869.7	34.46

Source: Ministry of Food Processing Industries, GOI, Annual Report 2008–09

The Indian food industry is projected to grow by US$ 100 billion to US$ 300 billion by 2015, according to a report by a leading industry body and Technopak. During the period, the share of processed food in terms of value is expected to increase from 43 per cent to 50 per cent.

D. Beverages

The Indian non-alcoholic drinks market was estimated at around US$ 4.43 billion in 2008 and is expected to grow at a compound aggregate growth rate (CAGR) of around 15 per cent during 2009-2012, according to a report published by market research firm RNCOS, titled "Indian Non-Alcoholic Drinks Forecast to 2012".

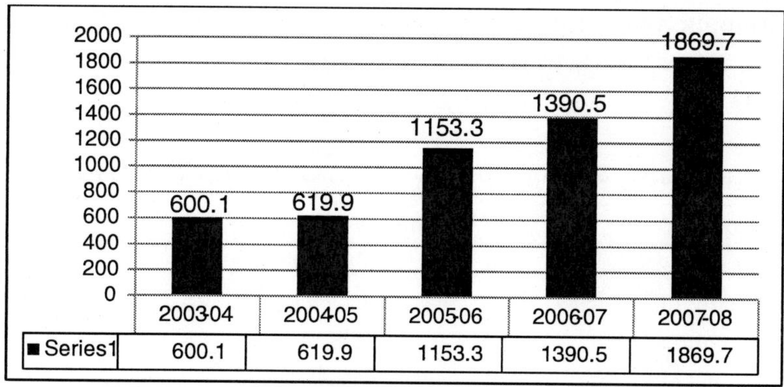

	2003-04	2004-05	2005-06	2006-07	2007-08
■ Series1	600.1	619.9	1153.3	1390.5	1869.7

Figure 24.1: *Exports of Processed Food (in US$ million)*

As per the report, the fruit/vegetable juice market will grow at a CAGR of around 30 per cent in value terms during 2009-2012, followed by the energy drinks segment which will grow at a CAGR of around 29 per cent during the same period.

France's Groupe Danone is merging its distribution operations in India with its probiotic drinks joint venture (JV) Yakult Danone India Pvt Ltd for better synergies.

Rural Indian households are reportedly spending more on consumer goods like durables, beverages and services than five years ago.

2. *Investment Trends & FDI Policy*

FDI in the Indian food processing industry is allowed under the automatic route (Entry of foreign/large players is restricted for a few food items) in agri products, milk and milk products, and marine and meat products, except the following (broadly) Proposals that require an industrial licence; and cases where foreign investment exceeds 24 per cent equity in units that manufacture items reserved for the small-scale industries.

(1) Proposals in which the foreign collaborator has a previous venture or tie-up in India, as on January 12, 2005.

(2) Proposals falling outside notified sectoral policy/caps or are generally restricted/prohibited.

(3) Repatriation of profits and capital permitted.

(4) Automatic approvals for foreign investment and technology transfer in most cases.

(5) Units based on agriproducts that are100 per cent export-oriented, are allowed to sell up to 50 per cent in domestic market.

(6) No import duty on capital goods and raw material for 100 per cent export-oriented units.

(7) Exemption of earnings from export activities from corporate taxes.

* Note: Entry of foreign/large players is restricted for a few food items.

TABLE 24.2

Cumulative FDI Inflows: Period: April 2000 to January 2010

SECTOR	*Amount of FDI inflows*	*Percentage of total*
Agricultural services	1,496.76	40.52
Food processing industries	1,018.97	27.60
Fermentation industries	767.72	20.79
Agricultural machinery	149.31	04.05
Vegetable oils and vanaspati	129.82	03.51
Tea and coffee	89.14	02.41
Sugar	41.68	01.12
Total	3,693.40	100.00

Source: "Fact Sheet on Foreign Direct Investment (FDI)", Department of Industrial Policy and Promotion website, www.dipp.nic.in, accessed 29 April 2010.

The Indian food processing industry needs at least US$ 35 billion fresh investments across sectors. This will enable the industry to create the projected nine million jobs, stability in food prices, reasonable returns to farmers and other stakeholders, and more importantly, to increase India's share in the world export market for processed foods from the current 2 per cent, stated the experts at Food Con 2011, organised by the Confederation of Indian Industry (CII).

A. The food processing industries in India has attracted foreign direct investment (FDI) worth US$ 1,273.96 million from April 2000 to June 2011, according to the data provided by Department of Industrial Policy and Promotion (DIPP).

B. South Korea has emerged as a major trade basket for India, according to H E Kim Joong Keun, Ambassador Extraordinary and Plenipotentiary, Republic of Korea in an interactive session organised by CII, Chandigarh. "Korean companies expect a great upsurge in the food processing industry especially in Punjab, it being the agrarian state of India," added JoongKeun.

C. Lowestoft firm Starfrost has won a contract worth £250,000 to supply specialised freezing equipment to a food processing plant in India. Starfrost will supply a quick freezing system for a 141-acre food processing facility in the Chittoor district of Andhra Pradesh in South-East India.

D. US-based McCormick & Co Inc, a leading spice maker, plans to invest about US$ 115 million in a joint venture (JV); it will form with Kohinoor Foods Ltd, a leading marketer of branded Basmati rice and other food products.

E. CavinKare Group has forayed into the confectionery segment in India with its liquid candy – Funfills. The organised confectionery market in India is estimated at approximately Rs 3,000 crore (US$ 635.32 million), as per Sanjay Sachdeva, the firm's Business Head, Foods & Snacks.

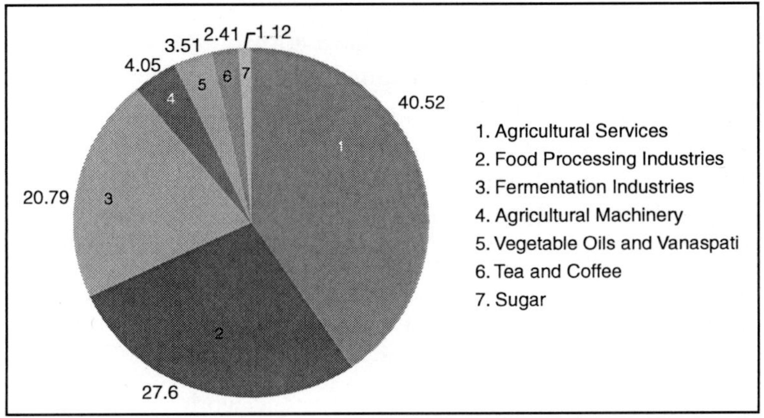

Figure 24.2: *Exports of Processed Food (in US$ million)*

3. Market Players and Strategies

- The ready-to-eat (RTE) segment stands at US$ 17 million to US$ 22 million and is growing rapidly at 30 per cent per annum, as per Asheesh Sharma, Marketing Manager, AgroTech Foods pointed out that according to the Ministry of Food Processing Industries
- Mother Dairy is not the only firm offering the new platter of deep-freezing snacks. From Godrej Tyson Foods' Yummiez brand to Canadian French fries giant McCain to India Equity Partners' newly-acquired Sumeru brand, a host of players are cooking up new food offerings
- The branded frozen foods category is estimated at Rs 1,000 crore (US$ 211.77 million), and industry players say it is now growing along at 20-25 per cent per annum
- Amul has been ranked as the first Indian Brand, in the list of the top 1,000 brands in Asia by The Campaign Magazine, published from Hong Kong and Singapore. Amul is also ranked the No.1 dairy brand, ahead of leading food and dairy brands of the Asian region, including Dutch Lady, Dumex and Magnolia
- Cooperative dairy giant Amul aims to expand its presence out of India. It is considering opening a processing facility in the US and subsequently in the European market. "There is a wide scope to produce and market other dairy products locally in US as brand Amul is preferred by Indians even overseas," as per Rahul Kumar, MD, Amul Dairy.

4. *Organized Retail Sector Emerging In India*

The study brings out several findings that have serious policy implications. The organized retail which accounts for 5 per cent of the total retail trade is poised to grow at an annual rate of around 11 per cent and is likely to touch business levels of 53,000 billion by 2020. Agri-food retailing accounts for 18 per cent of the organized retail today and is likely to have a lower share (12%) by 2020. The study has identified a few major impediments, especially structural, hampering the growth of organized retail. Direct sourcing by retailers from farmers is less prevalent though it is most desirable and in the interest of all stakeholders. India's agrarian culture and varied regional climate has made a significant contribution to the global food basket. Indian curries, mangoes, snacks and spices are known for their excellent quality worldwide. Globally, India holds the top position in the following food segments:

- India is the largest producer of milk in the world (108.5 million tonnes).
- India has the largest cattle population (294 million).
- It is the largest producer of mangoes in the world (15 million tonnes).
- It is also the largest producer of bananas (27 million tonnes).
- It occupies the second position in fruit (72.2 million tonnes) and vegetable production (133.5 million tonnes).
- It is the third-largest producer of fish in the world (7.6 million tonnes).

Agriculture is the mainstay of the Indian economy, providing employment to 52 per cent of its workforce. The agriculture and food processing sector plays an instrumental role in augmenting the growth of the economy, as it is an important source of raw material for the industrial sector.

- India is fast becoming the retail destination of the world. According to statistics, India has emerged as the leader in terms of retail opportunities. The retail market in India is anticipated to grow to 427 billion USD by the year 2012.
- The rapid growth of modern retailing, especially in food and groceries, calls for an organized cold chain infrastructure. India is one of the largest producers of fruits and vegetables in the world. Moreover, India's food market is valued at $70 billion, which is doubling every three years.
- The major factors fuelling this change are the
  - I. Increase in disposable income of the people,
  - II. Improving lifestyles,
  - III. Increasing international exposure and increasing awareness among the customers.
  - IV. India has a large middle class as well as youth population, which has contributed greatly to the retail phenomenon. The middle class

is considered to be a major potential customer group. The youth are perceived as trend setters and decision makers.

V. Tourist spending in India is increasing, which has also prompted the retail boom.

- The FDI inflow in this sector from 2005–06 was approximately US $74 million. Thus, investment opportunities in the food processing sector of India are immense. Not only does the country have the highest food production in the world, it also has a wide variety of crops, fruits, vegetables, livestock, and seafood.

5. *Government Initiatives and Duty Regime*

Some of the important policy measures and initiatives taken by the Government for the sector are:

- Most of the processed food items have been exempted from the purview of licensing under the Industries (Development & Regulation) Act, 1951, except items reserved for small-scale sector and alcoholic beverages;
- Food processing industries are included in the list of priority sector for bank lending in order to ensure easy availability of credit to them;
- Excise duty on ready to eat packaged foods, instant food mixes like dosa and idli mixes, aerated drinks, as well as on fruits and vegetables processing units, have been reduced.
- Excise duty on processed meat, fish and poultry products reduced from 8% to nil.
- Excise Duty of 16% on dairy machinery has been fully waived off.
- Excise duty reduction from 16 per cent to 8 per cent on a few more items including water purification devices, veneers and flush doors, sterile dressing pads, specified packaging material and breakfast cereals.
- Excise duty exemption on refrigeration equipment will enhance investments in the cold chain sector and help food and beverage sector. Higher standards ....making a difference for you.
- Customs duty on food processing machinery and their parts have been reduced from 7.5% to 5%.
- A large number of foreign collaborations have been approved.
- Up to a maximum of 24% foreign equity is allowed in SSI sector.
- Use of foreign brand names is now freely permitted.
- MRTP (Monopolies & Restrictive Trade Practices Act) rules and FERA (Foreign Exchange Regulation Act) regulations have been relaxed to encourage investment and expansion by large corporates.
- Most of the items can be freely imported and exported except for items in the negative lists for imports & exports. Capital goods are also freely importable, including second hand ones in the food processing sector.

- Units in Export Processing Zone / Free Trade Zone and 100% Export oriented units can retain 50% of foreign exchange receipts in foreign currency accounts.
- 50% of the production of Export Processing Zone / Free Trade Zone and 100% EOU units are saleable in domestic tariff area.
- All profits from export sales are completely free from corporate taxes. Profits from such exports are also exempt from Minimum Alternate Tax (MAT).
- The government has proposed for the establishment of mega food parks in different parts of the country, which will be run by a Special Purpose Vehicle created by all the stakeholders to create an integrated value chain from the farm gate to the consumer.
- Another strategic initiative taken in India is to establish cold chain facilities including refrigerated vans all over the country, to provide relief to the farmers, to enhance the shelf life of their product and retain its quality.

6. *Recommendations & Policy Implications*

The objectives of agro-processing programmes in India should be to:

- minimize product losses,
- add maximum value,
- achieve high quality standards,
- keep processing cost low,
- ensure that a fair share of added value goes to the Producer

National plan for improvement and extension of agro-processing technology at farm, traditional

(a) Small industry and modern industry levels should be prepared. The plan should take into account the diversity in resources and needs of different regions in the country. It should include programme details and implementation schedule for the first four or five years. The progress of plan implementation should be periodically reviewed to allow adjustments and corrective measures, and to develop programme details for the years beyond the period under review.

(b) Thrust areas for research and development should be identified and medium term research and development programme should be prepared and implemented to support the national plan for improvement and extension of agro-processing technology at different levels. Treatment and utilization of effluents from agro-processing industry should be include in R . D. programme.

(c) Emphasis should be put on the establishment of new agro-industrial plants in the production catchments to minimize transport cost, make use lower cost land and more abundant water supply, create employment opportunity in the rural sector and utilize process waste and by-products for feed, irrigation and manure.

(d) Infrastructure in the production catchments selected for agro-industrial development should be improved. Because of uncertain grid power supply to rural areas, decentralized power generation using locally available resources may be come an integral part of agro-industrial development. Similarly, if the raw materials and processed products are perishable or semi perishable in nature, cold chain will have to be established.

(e) The national plan should be provide for management of agro-industrial activities in the catchment area, both by private companies and individuals as well as cooperatives.

(f) Financial incentives and support should be provide on liberal scale to promote the modernization of agro-processing industry and for establishing new such industries in production catchments.

(g) Arrangements to supply market information to the farmer and agro-processor should be put in place.

References

Flavors of Incredible India, Ernst & Young, 2009.

Corporate Catalyst India (CCI) Survey Report, Press Releases.

KPMG Report on Food Processing and Agri Business.

Agriculture and Processed Food Products Export Development Authority (APEDA) articles.

Ministry of Food Processing Industries articles.

RNCOS Research Report, Department of Industrial Policy and Promotion (DIPP), Media Reports.

Cygnus Research, Oct 2008 www.ibef.org

OO

25

DEVELOPMENT OF PASTA WITH INCORPORATION OF ALOEVERA

Devendra Kumar Bhatt

Pasta is a generic term for foods made from an unleavened dough of wheat or buckwheat, flour and water, sometimes with other ingredients such as eggs and vegetable extracts. Pastas include noodles in various lengths, widths and shapes, and varieties that are filled with other ingredients like ravioli and tortellini. The word pasta is also used to refer to dishes in which pasta products are a primary ingredient. It is usually served with sauce. There are hundreds of different shapes of pasta with at least locally recognized names. Examples include spaghetti (thin rods), macaroni (tubes or cylinders), *fusilli* (swirls), and *lasagne* (sheets). Two other noodles, *gnocchi* and *spätzle*, are sometimes considered pasta. They are both traditional in parts of Italy.

It is generally accepted in United States that semolina from durum wheat is the material of choice for producing the highest quality pasta products. Pasta products can be processed from non-durum wheat farina or flour alone or blended with semolina. Pasta products refer to macaroni, sphegetti, noodles, each of which can be marketed in variety of shapes and sizes. Pasta is categorized in two basic styles: *dried* and *fresh*. Dried pasta made without eggs can be stored for up to two years under ideal conditions, while fresh pasta will keep for a couple of days in the fridge. Pasta is generally cooked by boiling. In flour and semolina for pasta, critical properties are protein content, gluten strength and gluten quality. Proteins are important in pasta because they form an insoluble network that traps swollen starch granules during cooking. This prevents disintegration of the pasta surface that makes it soft and sticky.

The pasta can be fortified with many fruits and vegetable powders because of the various nutritional and therapeutic properties. In the present investigation pasta was developed by fortification with aloe vera. The fortification of ginger, tulsi, aloe vera, carrot (spent) and soy protein isolate improves digestibility, nutrients availability and sensory quality of the product.

Pasta and Noodles Making Machine

The Pasta may be prepared by the pasta making machine P-12 (LaMon Ferrina, Italy). The machine structure is made up of entirely pure and unalterable stainless steel (AC), Polycarbonate (PC) and electrical motor is made up of Aluminium (Al), steel (AC), Copper (Cu) and Polyamide (PA)

Process of Making Pasta and Noodles

Any type of flour (wheat flour, rice flour) mixture can be used in the machine for making the dough. The dough must be needed with 30% of water (approximately)/ kg of wheat flour and other ingredients (ginger powder, aloevera gel, aonla, tulsi spent carrot, soya protein isolate in different percentage) in mixing chamber.

1. Firstly all the flours are weighed and sieved and then the following steps are applied.
2. Open the lid and fill mixing tank with the basic ingredients and water.
3. Close the transparent Plexiglass cover and secure with the relevant safety catch
4. Turn the master switch on for kneading.
5. The dough will be ready for extrusion after around 15 min.
6. Turn the selector knob from the kneading position to Extrusion mode.
7. Press operation button
8. Then begin to cut the dough with blade.
9. The cooling blower will come into the operation and partially dry the dough leaving the plate.

Aloevera Fortified Pasta

Raw aloe vera was procured from local surroundings and adjoin villages of Jhansi, were washed in running water to remove adhered dust, dirt and mucilaginous Aloevera was cut and dried to powder. Wheat flour-aloe vera blends were prepared separately by manual mixing. Aloe vera was added in wheat flour in various percentage i.e. 1 , 2, 3, 4 and 5. Blends were put into the pasta maker for kneading with the addition of 250 ml water.

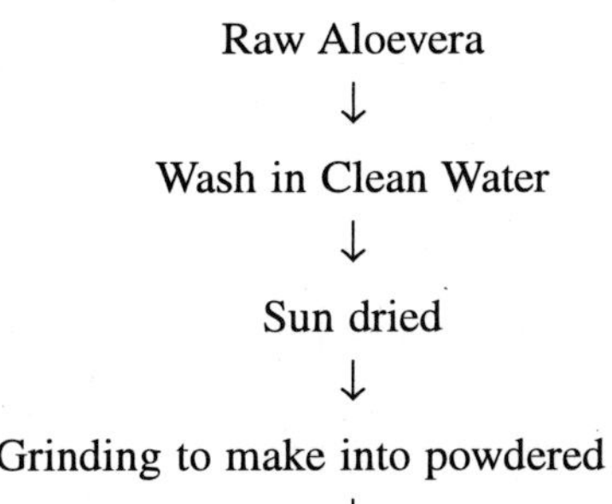

Addition of Wheat flour-Aloevera powder with water

↓

Kneading at room temperature

↓

Extruded

↓

Drying of fresh pasta and noodles in tray

↓

Extruded and dried pasta and noodles are collected

Figure 25.1: *Preparation of Aloevera Fortified Pasta and Noodles*

Development of Pasta and Noodles (Control)

Optimization of time

Firstly all the flours are weighed and sieved and then open the lid and fill mixing tank with the basic ingredients and water. Close the transparent Plexiglass cover and secure with the relevant safety catch. Turn the master switch on for kneading. The dough will be ready for extrusion after around 15 min. Turn the selector knob from the kneading position to extrusion mode. Press operation button. Then begin to cut the dough with blade. The cooling blower will come into the operation and partially dry the dough leaving the plate.

Statistical Analysis

Data was subjected to statistical analysis using ANOVA with statistical software (Systat 11 soft ware).

TABLE 25.1

Chemical Constituents of Aloevera Powder

Constituents	*Quantity*
Moisture (%)	4.5
Protein %	4.34
Fat (%)	2.12
Ash (%)	14.97
Carbohydrate (%)	74.07
Crude fibre (%)	17.41

Preparation of Wheat Flour Aloevera Pasta

Standardization of the Ratio of Wheat Flour with Aloevera

Wheat flour and Aloe Vera were blended in different ratios for the preparation of wheat flour aloe vera pasta. Aloe Vera was mixed in a ratio of 1, 2, 3, 4, and 5 percent. That is the ratio of wheat flour to aloe Vera mixed is 99:01, 98:02, 97:03, 96:04and 95:05 percent ratios. The blended pasta was prepared. The wheat flour aloe Vera pasta was evaluated by a sensory panel of 10 members by using 9 point hedonic scale, The sensory scored were 6.45, 6.47, 6.62, and 7.08 percent for 99:01, 98:02, 97:03, 96:04 and 95:05 ratio(wheat flour : Aloe Vera) respectively. The highest sensory score of 7.08percent was found of 95:05 ratio refined wheat flour and aloe Vera (table 4.22) The sensory score of 95:05 ratio was significantly different (P£0.05) in comparison to other combination (CD = 0.083).

TABLE 25.2

Standardization of the Ratio of Wheat Flour with Aloevera

Treatment	*Color*	*Body & Texture*	*Flavour/ Taste*	*Overall Accept-ability*
WF:AV(99:.01)	6.08	6.25	6.96	6.45
WF:AV(98:02)	6.16	6.27	7.4	6.47
WF:AV(97:03)	6.21	6.39	7.16	6.58
WF:AV(69:04)	6.34	6.42	7.24	6.62
WF:AV(95:05)	6.72	7.09	7.32	7.08
CD (5%)	0.209	0.098	0.093	0.083
P value	0.019	0.000	0.004	0.003

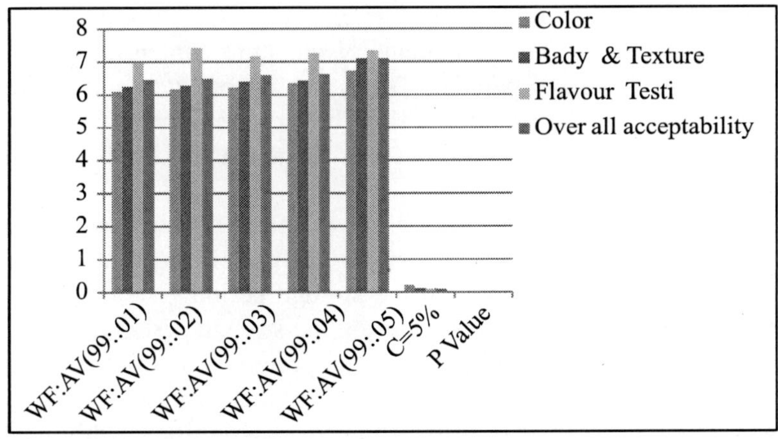

Figure 25.2: *Standardization of the Ratio of Wheat Flour with Aloevera Powder*

Storage Study

The selected wheat flour Aloe Vera pasta and plain pasta was stored for the storage study. The wheat flour Aloe Vera pasta and plain pasta was packed in air tight zip pouches and stored in room temp. for six months. Physico- chemical characteristic were evaluated at an interval of 0, 1, 2,3,4,5, and 6 months at room temp. (16-35°C). Organoleptic, textural and microbiological changes were observed.

Physico-chemical Characteristics

The chemical constituents present in wheat flour aloe Vera pasta influence the nutritional and storage quality of the product. The plant has been presented as a constituent of many phytochemicals, vitamins, nutrients and anti-nutrients found in foods **(Maenthalsong *et al.*,2007)**. Wheat flour Aloe Vera pasta was analyzed for proximate composition as per the approved methods. The moisture content was analyzed by over drying method, protein by micro kjeldahl method, fat by soxhlet extraction method, total ash by muffle furnish and crude fiber by AOAC, (1998). All the constituents were analyzed at the end of 0,5,10, and 15 days of the storage at room temp. (16-35°C).

Physico-chemical Changes

The product was stored for six months and the changes in moisture, protein, fat, ash, crude fiber and carbohydrate, were observed. The results of the observations are presented in table 3.

The moisture content of the wheat flour Aloe Vera pasta on zero days was recorded 6.39 percent. A non-significant change in the moisture content of wheat flour aloe vera pasta was noted throughout the storage period. These changes in the moisture content of pasta were assessed during the storage period of 6 months with an interval of 1 month (table 3). The initial moisture content of pasta was 6.39 percent after 1 month of storage. The change of moisture content showed a non-significant difference. The percentage of moisture content of pasta after two months of storage was 6.48 percent. There was a non-significant difference in the moisture content of the product during the storage period. The difference of moisture content of pasta after 3 months of storage was 6.52 percent and showed a non-significant difference.

The protein content of wheat flour aloe vera pasta on zero days was recorded 11.89 percent. A non-significant change in the protein content of wheat flour aloe vera pasta was noted throughout the storage period. The change in protein content of pasta samples was assessed during the storage period of 6 months with an interval of 1 month (table 25.3). The initial protein content of pasta was 11.88 percent after 1 month of storage. The protein content of pasta after 6 months of storage was 11.81 percent. The result showed non-significant difference in the protein content of pasta during 6 months of storage.

TABLE 25.3

Changes in Chemical Constituents of Wheat Flour Aloe Vera Pasta during Storage

Storage period (Months)	*Moisture (%)*	*Protein (%)*	*Fat (%)*	*Ash (%)*	*Crude fiber (%)*	*Carbohydrate (%)*
0	6.39	11.89	3.86	1.2	1.53	75.89
1	6.41	11.88	3.84	1.118	1.52	75.88
2	6.48	11.87	3.84	1.16	1.52	75.56
3	6.52	11.87	3.81	1.16	1.51	75.84
4	6.58	11.86	3.79	1.15	1.51	75.81
5	6.65	11.86	3.79	1.13	1.50	75.76
6	6.75	11.81	3.78	1.12	1.49	75.71
CD 5%)	0.380	0.777	0.006	0.042	0.189	0.294
P value	0.099	1.000	0.000	0.999	1.000	0.998

The fat content of wheat flour aloe vera pasta on zero days was recorded 3.86 percent. The fat content of pasta after 6 months of storage was 3.38 percent. The difference of fat content showed a significant difference.

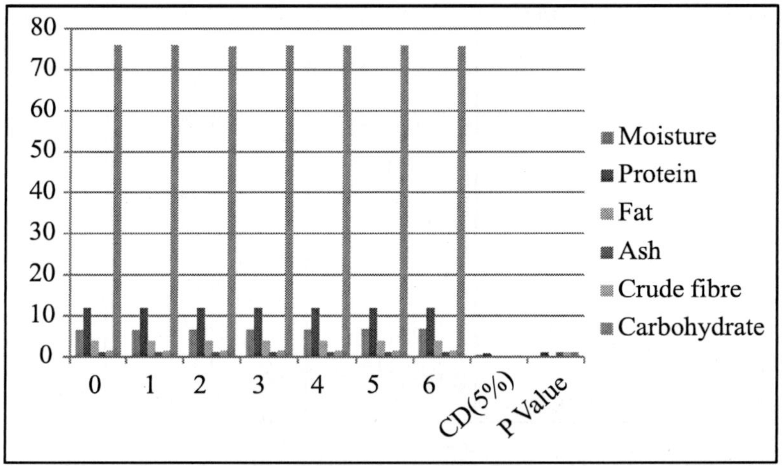

Figure 25.3: *Changes in Chemical Constituents of Wheat Flour Aloe Vera Pasta during Storage*

The total ash content of wheat flour aloe vera pasta was 1.2 percent. The crude fiber content of the wheat flour aloe vera pasta on zero days was 1.53 percent. The crude fibre content of pasta after 6 months of storage was decreased to the level of 1.49 percent from the initial 1.52 percent. The result showed non-significant difference.

The carbohydrate content of wheat flour aloe vera pasta on zero days was recorded 75.87 percent. The percentage of carbohydrate content of pasta after 6 months of storage was decreased to the level of 75.71 percent from the initial 75.89 percent. The result showed carbohydrate content of pasta was non-significantly different throughout the storage.

Organoleptic Evaluation

The acceptability of wheat flour Aloe Vera pasta was evaluated by 10 member panel. The sensory feed back of wheat flour aloe Vera pasta was taken on a 9 point hedonic scale from panel members on the different quality parameters (color, texture taste flavor, over all acceptability).

Changes in the Organoleptic Qualities during Storage

The final product was stored for the determination of storage quality. The effect of storage on the organoleptic qualities of wheat flour aloe vera pasta was assessed during a storage period of 6 months with an interval of 1 month (table 4)

TABLE 25.4

Changes in Sensory Attributes in Wheat Flour Aloevera Pasta during Storage

(Storage period) Months	*Colour*	*Texture*	*Taste /Flavour*	*Over all acceptability*
0	9.18	9.16	9.27	9.20
1	9.10	9.08	9.15	9.10
2	9.06	9.03	9.04	9.04
3	8.96	9.99	8.89	8.94
4	8.85	8.80	8.74	8.76
5	8.81	8.81	8.60	8.69
6	8.73	8.77	8.50	8.62
CD (5%)	1.137	0.114	0.126	0.140
P value	0.010	0.010	0.010	0.010

The sensory score for over all acceptability of wheat flour aloe vera pasta on zero days was 9.20. During the storage period the over all acceptability score showed a declining trend in pasta. The changes in the over all acceptability score of pasta sample were assessed during the storage period of 6 months with an interval of 1 month (table 4). The initial sensory score for over all acceptability of pasta was 9.20, which was decreased to 9.10 after 1 month. The change in the over all acceptability score of pasta showed significant difference (P£0.05). The sensory score for over all acceptability of pasta after 2 months of storage was decreased to 9.04 and showed a significant difference (P£0.05) in the pasta product during the storage period.

The sensory score for over all acceptability of pasta after 3 months of storage was decreased to 8.94 and showed a regular decline in the over all acceptability score of pasta product and showed a significant difference (P£0.05). The sensory score for over all acceptability of pasta after 4 months of storage was decreased to 8.76 and showed a significant difference (P£0.05). The sensory score for over all acceptability of pasta after 5 months of storage was decreased to 8.69 and showed regular decline in over ail acceptability of the pasta product and showed a significant difference (P£0.05).

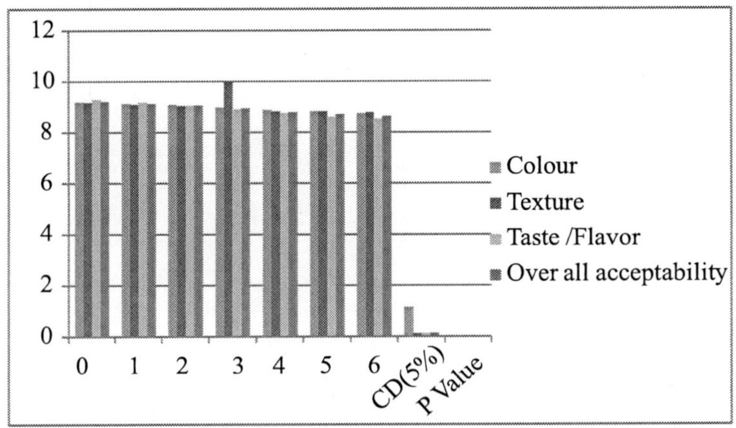

Figure 25.4: *Changes in Chemical Constituents of Wheat Flour Aloe Vera Pasta during Storage*

The sensory score for over all acceptability of pasta after 4 months of storage was decreased to 8.76 and showed a significant difference (P£0.05). The sensory score for over all acceptability of pasta after 5 months of storage was decreased to 8.69 and showed regular decline in over all acceptability of the pasta product with a significant difference (P£0.05). The sensory score for over all acceptability of pasta after 6 months was decreased to 8.62 from initial score 9.20. The result showed the change in the over al acceptability score was significant at 5 percent level of significance (CD = 0.040).

Textural Analysis

The textural characteristics of the pasta and noodle samples were determined to evaluate the changes in the hardness characteristic of the pasta and noodle. Many instrumental measurements have been reported for the evaluation of pasta texture **(Matsuo and Irvine 1969, 1971; Walsh 1971; Matsuo *et al.,* 1972; Voisey and Larmond 1973; Dexter *et al.,* 1983a,b, 1985; D'Egidio et al 1993; Edwards *et al.,* 1993, 1995; Malcolmson et al 1993; Guan and Seib 1994; Kovacs et al 1995; Smewing 1997; Gonzalez *et al.,* 2000).**

TABLE 25.5

Changes in Texture of Pasta and Noodle Sample during Storage.

Hardness test.

(Storage days) *Treatment*	*Wheat flour Aloevera* *Pasta and noodle*
0 Days	9.33
1 Month	9.47
2 Month	9.53
3 Month	8.73
4 Month	8.53
5 Month	8.53
6 Month	8.00
CD	0.338
P Value	0.035

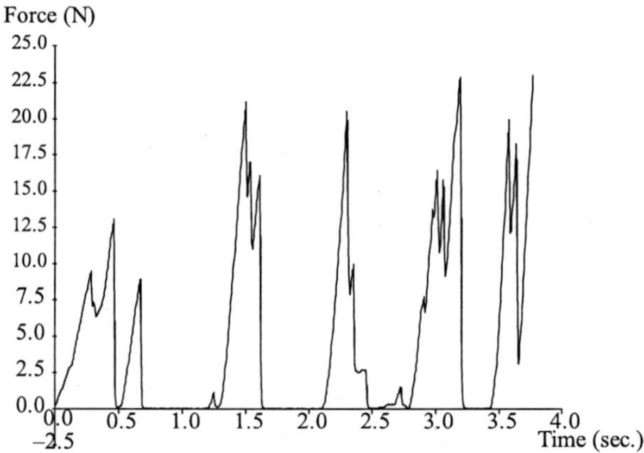

Wheat Flour Aloevera Pasta

Figure 25.5: *Hardness of Blended Pasta and Noodle during Storage*

The hardness of the pasta and noodle were gradually decreases during the storage period. The changes in the hardness strength of the pasta and noodle samples were assessed during the storage period of 6 month with an interval of 1 month (table 5).

Microbiological Analysis of Pasta and Noodles

The total plate count was nil at initial stage of storage. The effect of storage on the quality of pasta and noodles was assessed during the storage period of 6 months with an interval of 1 month (Table 6). Total plate counts were not observed upto 4 months of storage.

TABLE 25.6

Microbial Count in Pasta Samples During Storage

Storage period	*Total Plate Count*	*Yeast & Mould Count*
	0	*0*
0 days	0	0
1 Month	0	0
2 Month	0	0
3 Month	0	0
4 Month	1.0×10^2	1.0×10^2
5 Month	2.0×10^1	2.0×10^1
6 Month	0.435	0.435
CD (5%)	0.01	0.00

Conclusion

Pasta is a very commonly used food items these days. The incorporation of aloevera will make the product a health food. The product developed by adding different percentages of aloevera was found acceptable and the storage study showed that even after six months the product was good and acceptable. The product was good in colour and texture.

References

Anderson, Y., Hedlund, B., Jonsson, L., and Svensson, S.,1981. Extrusion cooking of a high fiber cereal product with crisp bread character. *Cereal Chem.* 58-370.

APHA, 1992. Recommended methods for the microbiological examination of foods. American Public Health Association, Inc. New York.

Asp, N.G. and Bjorck, I., 1984. The effect of extrusion cooking on nutritional value. *Eds Elsevier Applied Science Publ.* London. 162.

Coats, B.C.,1994. Method of processing stabilized *Aloe vera* gel obtained from the whole *Aloe vera* leaf. US Patent, 5, 356, 811.

Davis, R.H.; Kabbani, J.M.; Maro N.I., 1987: Aloe vera and wound healing. *J Am Podiatr Med Assoc.* 78 (2) 165-9.

Edwards, N.M., Biliaderis, C.G. and Dexter, J.E.,2006. Textural Charac- teristics of whole wheat pasta and pasta containing non starch polysachharides. *J. Fd. Sci.* 60 (6) : 321-324.

Espin, J.C., Conesa, Garcia, M.T., Barberan, Tomas, F.A., 2007. Nutraceuticals: facts and fiction. *Phytochemistry.* 68(22-24):2986-3008.

Fapojuwo, O.O., Maga, J.A. and Jansen, G.R.,1987. Effect of extrusion cooking on *in-vitro* protein digestibility of sorghum. J. Food Sci. 52, 218.

Voisey, P.W., and Larmond, E. 1973. Exploratory evaluation of instru- mental techniques for measuring some textural characteristics of cooked spaghetti. *Cereal Sci. Today* 18:126-133,142-143.

Wolf A., and Denninger, K.,2004. Food containing *Aloe vera* gel br *Aloe vera* powder and use of *Aloe vera* gel br *Aloe vera* powder in manufacture of foods Patent. DE 103 22025.

OO

ISO 22000:2005– Food Safety Management Systems

Aswani Duggal

Introduction

Concern for the supply of food that is safe for the consumer has increased over the years. Rising liberalization of agro-industrial markets and the world-wide integration of food supply chains require new approaches and systems for assuring food safety. Food processors and retailers are sourcing their ingredients worldwide and it can be hard to track the region let alone the producer of the ingredient. For these reasons and others, global retailers, distributors, food manufacturers and food service companies are now concerned more about the safety of their food supply chain than ever before.

At present, concern over food safety is at an all-time high. With each food "scare" reported from banned chemicals (like melamine) in multiple products to links between animal and human diseases (the latest being swine flu)- consumer concern grows. In response, the public and the private sector have developed new process standards and require suppliers of food products to follow them. Both, the market and legislations in importing countries demand for comprehensive and transparent schemes reaching "from farm to fork". Organizations in the food sector will need to manage risk, demonstrate good corporate responsibility and meet legal requirements if they are to remain competitive, protect their reputation and enhance their brand.

Need for Food Safety Standard Systems

Food safety is related to the presence of food borne hazards in food at the point of consumption. Food reaches to consumers via supply chains that may link many different types of organizations. One weak link can result in unsafe food that is dangerous to health. As food safety hazards can occur the food chain at any stage, adequate control throughout the supply chain is essential. Therefore food safety is a joint responsibility of all organizations with in the food chain including, producers, manufactures, transport & storage operators, sub contractors, retail and food service outlets and service providers.

Recent studies have shown that there is significant increase of illness caused by infected food in both developed and developing countries which give rise to considerable economic costs besides being health by all types of organizations within the food chain.

Even the WTO Agreement on Sanitary and Phytosanitary Measures (SPS) sets important requirements for adoption and implementation of food safety and quality and recognizes the standards, guidelines and recommendations determined by the Codex Alimentarius Commission (CAC). Excellence in food quality and safety had taken a tangible form with the advent of ISO 9000 Quality Management System and Hazard Analysis Critical Control Point (HACCP) standards. ISO 9000 encompasses all the activities of a company to ensure that it meets its quality objectives, while HACCP is directed towards ensuring food safety. The ISO 9000 standards were brought by the International Organization for Standardization (ISO) and the HACCP standards by the Codex Alimentarius Commission (CAC). These standards have assumed importance worldwide both as an essential requirement to tap the market potential and as a marketable feature of the company.

The concern for safe food led to the development of numerous food safety standards both national/international, private/public and voluntary/mandatory. The growing number of standards for food safety management led to confusion. Consequently, there was a need for international harmonization and ISO aimed to meet this need with ISO 22000:2005.

Tailor-made Approach

ISO 22000 has been designed with flexibility to enable a tailor-made approach to food safety for all segments of the food chain. It does not take a "one size fits all approach", since the standards and procedures required for high risk areas in one food sector may not be appropriate in another. For this reason, unlike other schemes, the standards does not provide a checklist methodology if a company seeds certification to ISO 22000, it needs to follow local and export market laws, as well as implementing customer's requests. ISO 22000 requires that industry targets each specific type of food product adequately according to its needs

Effectiveness through Communication

Through the development of one system that crosses all food sector branches and national borders, food safety is strengthened by the harmonization of working procedures. This is one of the fundamental rationales behind the ISO 22000 standard. If everyone uses the same methods and language, the systems effectiveness improves, increasing food safety, reducing the risk of critical errors and misunderstandings and maximizing the use of resources. ISO 22000 an be applied to all types of organization within the food chain, ranging from feed producers, primary producers, food manufactures, transport and storage operators and subcontractors to retail and food service outlets organizations such as producers of equipment packaging material, cleaning agents, additives and ingredients.

Making it Safe to Eat

ISO 22000 combines generally recognized key elements to ensure along the food chain.

Interactive Communication

Clear communication along the food chain is essential to ensure that all relevant food safety hazards are identified and adequately controlled at each step. This implies communication of an organization's needs to the other organizations both upstream and downstream in the food chain. Communication with customers and suppliers, based on the information generated through systematic hazard analysis, will also assist in meeting customer and supplier's requirements in terms of feasibility, need and impact on the end product.

System Management

The most effective food safety systems are designed, operated and updated within the framework of a structured management system and incorporated into the overall management activities of the organization. This provides maximum benefit for the organization and interested parties. ISO 22000 is aligned with the requirements of ISO 9001:2000 in order to enhance the compatibility of the two standards, and to ease their integrated implementation.

Hazard control

ISO 22000 combines the Hazard Analysis and Critical Control Point (HACCP) principles and application steps developed by the Codex Alimentarius, with prerequisite programmes. It uses a hazard analysis to determine the strategy for hazard control.

Advantages for the Food Industry

Organizations implementing the standard will benefit from:

- Organized and targeted communication among trade partners;
- Optimization of resources (internally and along the food chain);
- Improved documentation;
- Better planning, less post process verification;
- More efficient and dynamic control of food safety hazards;
- All control measures subjected to hazard analysis;
- Systematic management of prerequisite programmes;

- Wide application because it is focused on end results;
- Valid basis for taking decisions;
- Increased due diligence;
- Control focused on what is necessary ; and
- Saving resources by reducing overlapping system audits.

ISO 22000 can be applied to all types of organization with the food chain

ISO 22000:2005, Food safety management systems–Requirements for any organization in the food chain

ISO 22000 Key Elements

- ***Involvement of the Management Team***

Food safety is not just something to be handled by the quality department. It is a top management issue. ISO 22000 focuses on the involvement of the management team, which has to develop overall policy.

Communication: As food safety hazards may be introduced at any stage of the food chain, interactive communication both upstream and downstream is essential. In addition, internal communication is a key element to avoiding misunderstandings and minimizing risks. A common vocabulary is a great help in this connection.- The HACCP (Hazard Analysis & critical control point) principles ISO 22000 combines the recognized HACCP principles with prerequisite programmes. The hazard analysis determines a strategy and the prerequisite programmes set up an action plan.

- ***System Management***

ISO 22000 relies on a structured management system based on relevant parts of ISO 9001. It is possible to integrate them into one management system together with ISO 14001.

The Development Team

ISO 22000 was developed by food safety experts from Australia, Canada, Belgium, Denmark, France, Germany, Greece, Ireland, Japan the Netherlands, Poland, Sweden, Switzerland, Thailand, United Kingdom, USA, Venezuela, Vietnam and liaison bodies such as the Confederation of Food and Drink Industries of the European Union (CIAA), amongst others.

Food You Can Trust

Other stakeholders will benefit from:

- Confidence that the organizations implementing ISO 22000 have the ability to identify and control food safety hazards. The
- Standard adds value because it:

- Is an auditable standard with clear requirements;
- Is internationally accepted;
- Is a publicly available scheme, not a proprietary one, that can be used by all;
- Integrates and harmonizes various existing national and industry- based certification schemes;
- Addresses a desire for harmonization from the food processing industries concerning food safety;
- Is aligned with ISO 9001: 2000, Quality management systems- Requirements and ISO 14001: 2004, Environmental management systems- Requirements with guidance for use and the Occupation Health and Safety Assessment Series (OHSAS) and can also incorporate retailers' standards; and contributes to a better understanding and further development of HACCP.

A Complete Menu of Standards

ISO 22000 is supported by a complete set of standards that reinforce its implementation.

- ***Audit and Certification***

To increase the acceptance of ISO 22000 and ensure that accredited certification programmes are implemented in a professional and trustworthy manner, the technical specification: ISO/TS 22003, Food safety management systems- Requirements for bodies providing audit and certification of food safety management systems, was published in 2007 it provides the necessary information.

And confidence on how the certification of an organization's food safety management system has been conducted. This technical specification offers harmonized guidance of the accreditation of certification bodies, and defines the applicable rules for the audit of a food safety management system compliant with ISO 22000.

Applying Food Safety

ISO/TS 22004:22005, Food safety management systems- Guidance on the application of ISO 22000:2005, was published in 2005. It provides guidelines on implementing ISO 22000, with particular emphasis on good examples. In 2007, ISO and the International Trade Centre collaborated to produce the book, ISO 22000 Food safety Management Systems- An easy to use checklist for small business- Are you ready? It explains how small and medium- sized enterprises can use and implement.

ISO 22000

Tracing the Feed and Food Chain

A new standard was published in June 2007 on traceability: ISO 22005:2007, Traceability in the feed and food chain- General principles and basic requirements for system design and implementation. It is a useful amplification of the reference in ISO 22000 to traceability as an important component of food safety. ISO 22005 is intended for organizations operating or cooperating at any point in the feed and food chain. It does not contain any reference to certification nor is it combined with other standards Instead, the possibility of certification is left to the user's discretion. However, the standard requires that organizations carry out monitoring, internal audit and reviews to assess the effectiveness of the system. Other standards in the ISO 2000 family are being developed. They include a project for quality management for farmers, and a standard dealing radiation of foods.

Benefits of Building Food Safety

Food producers in all parts of the food chain around the world have adopted ISO 22000 as a new global food safety standard. Still many small and medium-sized companies are waiting for the position of the three main market drivers: multinational food companies, authorities and retailers. Some of the largest multinational food companies have been very positive about implementing the standard for themselves and their suppliers. Authorities in some countries plan to let certified companies benefit from less frequent controls, and perhaps consider an outsourcing of public control. The front runners in the work are authorities in Denmark and France. Retailers in Belgium and Denmark are already certified with ISO 22000, but greater communication on the benefits for retailers is still needed.

Driving Forces for the Future

It is important to create a common understanding of the benefits of building a food safety management system based on ISO 22000, for both manufacturing companies and retailers. This is being addressed by ISO/TC 34, Food products and food sector organizations for retailers, food producing companies and international certification bodies, among whom a bridging process has been initiated.

ISO 22000 contributes to a better understanding of HACCP

The ultimate objective is to put safe food on the tables of the consumer.

The driving forces for the future development of ISO 22000's implementation are to a large extent, retailers, important food producers and national authorities. Future dialogue among them is needed. Authorities have to define food safety objectives for hazards like salmonella, coli bacteria and campylobacter, and companies need to incorporate control measures that ensure these acceptable levels are met. Another challenge for the food industry is to manoeuvre between authorities, clients and internal company demands on how to make safe foods.

ISO 22000 provides guidance on this never-ending question and now the ISO 22000 series provides a set of tools for use by all interested parties. These include consultants, certification bodies, accreditation bodies, public authorities and all players in the food supply chain, from farmer to retailer. The ultimate objective is to put safe food on the tables of the consumer- you and me.

Eight Keys to Successful ISO 22000 Implementation

Practical Steps Forward

The logic behind management standards such as ISO 9001, ISO 14001 or ISO 22000 is to establish a series of actions to be implemented prior to building your system. These actions are the direct responsibility of management, and include:

- An initial analysis of customer needs, expectations and requirements;
- Identification (or confirmation) of regulatory requirements;
- Definition of a policy supported by measurable objectives.
- Planning of actions and resources needed to meet your objectives. Once these steps have been accomplished, management should define how it will control FSMS improvement.

This is the "management system" phase, including:

- Analysis of the individual results obtained;
- Evaluation of combined results;
- Formulation of recommendations for improvement.

On the basis of these findings, we propose a method of building a certifiable ISO 22000: 2005 – based system based on eight keys to success. To view this in a dynamic way, imagine competing in a long distance running race over hurdles. The interested parties, stakeholders in our FSMS, will be our supporters. They will keep a close watch on how we perform and will certainly encourage us to do better.

1st Key Element

How do we meet basic food safety requirements?

This is the staring signal and determines how your regulatory observance programme should be conducted. IT also provides the opportunity to see how your food safety team defines, approves, communicates, implements and monitors the prerequisite programmes (PRP). The general principles of food hygiene establish solid foundations which, if necessary, should be used jointly with guides to good hygiene practice. These principles apply to the entire food chain from primary production to consumption, and specify the hygiene controls that must be carried out at each stage. Therefore, this first key element addresses the review and diagnosis of such food safety requirements.

Purpose of 1st Key Element

The ultimate purpose here is to identify all the basic conditions and activities necessary to maintain, on a permanent basis, a hygienic environment during the production, handling, storage and provision of safe food or end-products for human consumption. Basic entry data to help develop these PRPs can include:

- Codes of Practice of the Codex Alimentarius:
- applicable and statutory requirements;
- good hygiene practice guides;
- contractual requirements with customers;
- trade practices and customs.

The organization must:

- identify the PRPs to be observed;
- implement PRPs effectively;
- ensure the permanent implementation of its PRPs. Management should be able to develop, formalize and communicate its policy to all interested stakeholders, based on this analysis.

2nd Key

How safe is our product?

ISO 22000:2005 requires that the organization use a dynamic and systematic approach to developing a FSMS, by

- implementing and monitoring planned activities, maintaining and verifying effective control measures;
- updating information; and
- taking appropriate action in the event of non-conformities. These guidelines can be found in the HACCP (Hazard Analysis and Critical Control Point) plan section of the standard. However, the organization should first address the steps preceding hazard analysis by:
- assembling an HACCP team with multidisciplinary skills;
- Producing a product description
- Defining the intended product use;
- Drawing up a flow diagram and on- site confirmation. And then proceed with the following steps:
- hazard identification and assessment (biological, physical or chemical);
- Selection and assessment of control measures to be aligned with a CCP (Critical Control Point) and /or an operational PRP;
- establishing and managing an HACCP plan and /or operational PRPs.

Purpose of 2nd Key Element

Clause 7 of ISO 22000:2005 addresses the phases of safe product planning and realization (HACCP method), while clause 8 deals with the phases of verification and action. System maintenance and improvement are addressed throughout the planning validation, monitoring, verification and updating cycles. This step-by step key will then enable the organization to efficiently implement and/or review its HACCP plan by incorporating the new ISO 22000:2005 concepts.

3rd Key

What are the steps to product safety?

Your top management should define its overall intentions and direction in food safety to enable measurable food safety objectives to be established. Establishing a safety policy requires the full commitment of all, starting with a management strongly committed to:

- mobilizing personnel so they feel involved and willing to comply with instructions;
- encouraging the achievement of objectives;
- creating favourable conditions for team work;
- communicating to personnel the importance of its ISO 22000:2005- based project, and raising awareness of food safety. Evidence of management commitment can be seen from awareness and leadership initiatives (attendance at meetings, training sessions, investment plans, etc.) intended to help develop and implement the FSMS.

Purpose of 3rd Key Element

The food safety policy should normally be consistent with the organization's overall policy, and provides a framework for setting objectives. ISO 22000:2005 requires that an organization analyze its position in the food chain to asses the degree of upstream and downstream control, and adjust its policy accordingly, Management should analyze:

- its main customer/regulatory and statutory requirements (see Ist key element);
- its main organizational weaknesses regarding food safety;
- the main environmental changes necessary. Management should be able to develop, formalize and communicate its policy to all interested stakeholders, based on this analysis.

4th Key

How should we plan the FSMS?

To achieve the objectives and ensure end-product safety, the process of system planning will require responses to the following questions:

- What do we need to do and how? In what order? Within what deadlines?

- Who is responsible? Who are the authorities?
- How should we build the documentation system?
- What physical and human resources should be deployed?
- How should we measure the work in progress?

Purpose of 4th Key Element

Means should be matched to objectives. IT is particularly important to structure the various actions and resources (notably budget) in accordance with a schedule. Good forward planning enables your organization to anticipate and handle any eventualities, while controlling impacts on food safety.

5th Key

What do we need for effective FSMS implementation?

Controlling food safety through an HACCP plan inevitably implies using technical resources and appropriate operational activities. At this stage, the HACCP plan should be well established. The means and activities that will make it function properly should now be identified. Such identification can include regulatory provisions- e.g., food traceability – and consensus- based or voluntary provisions, such as control of non-conformities and metrology of measuring equipment. Of course, each manufacturer is free to add any other means or activities it deems necessary or appropriate to those that have already been deter- mined.

Purpose of 5th Key Element

This key element addresses the implementation of an effective traceability system, supplemented by measuring, calibration and monitoring equipment to ensure that non-conformities do not occur.

6th Key

What should we do in the event of an accident?

Unfortunately, accidents can happen in the food-processing industry. Your company must have a system in place to respond to emergency situations, such as food safety alerts, outbreaks of food poisoning, process malfunctions, equipment failures, product withdrawals and recalls.

Purpose of 6th Key Element

It is essential to be in a position to handle all emergencies in the organization and avoid the internal or external impacts of such non-conformities.

7th key

How do we ensure effective FSMS implementation?

This element highlights verification of your FSMS effectiveness, and is covered in step 11 of the HACCP plan: Establish verification procedures. IT requires your verification activities to be planned to ensure that the actions taken been implemented and are effective.

Purpose of 7th Key Element

The verification process creates confidence that the FSMS you have implemented has been properly evaluated and is relevant. The organization must provide evidence of FSMS effectiveness in terms of the level to which planned activities have been realized, and the extent to which expected results have been achieved.

8th Key

How should we build on experience to improve the FSMS?

This is the continual improvement loop of the system. Such improvement is possible only after the analysis of data. You should take account of the results of verification activities before making decisions during the management review. This review should enable you to identify new resource needs, generate improvement actions and decide on the appropriateness of your policy.

Purpose of 8th Key Element

To achieve optimal food quality and safety, organizations must adopt a continual improvement approach, i.e., involve all parties in a daily quest for effectiveness. This key element determines the requirements that must be fulfilled to establish a truly dynamic system. To conclude the athletics analogy, successful ISO 22000: 2005- based FSMS implementation requires continual effort lap after lap. Indeed, the pursuit of continual improvement, particularly in food safety, demands that you jump over increasingly higher hurdles.

References

British Retail Consortium website, http://www.brc.org.uk/standards / index.htm

Global Food Safety Initiative System, http://www.ciesnet.com/pfiles/programmes/ foodsafety/ GFSI

ISO 22000 Standard http://www.iso.org/iso.

SQF Institute website, Safe Quality Food,www.sqfi.com

The Agrifood Division of Sai Global, http://www.efsis.com/

The International Food Standard web site, http://www.food-care.info/

World Food Safety Organization website, www.worldfoodsafety.org

ISO Management Systems www.iso.org/ims

27

Studies on Utilization of Buttermilk in Cookies Making

Neha Chaudhary, Silky and Dheer Singh

Introduction

Milk and milk products are traditional components of bakery items and are used for their several desired functions in food products .They improve moisture absorption capacity of dough's; facilitate dough handling ;prevent rapid and excessive acidification by enhancing buffer capacity during fermentation and; allow better control of amylase activity; facilitate moisture transfer during gelation of starch; minimize the effect over mixing; enhance flavor development and crust colour; improving toasting characteristics; strengthen crumb structure and texture; improve moisture retention and retard staling process and enhance nutritional value.

An increased trend in the use of skim milk, buttermilk and whey in western countries for enriching milk products and as a basis for manufacturing other food products has been observed. These byproducts are least profitable. The disposal treatment and utilization of dairy waste is subject of growing importance in view of the increasing emphasis on the environmental and economic aspects. Draining of these byproducts not only cause loss of valuable solids, but also result in high biological oxygen demand (BOD) in waste water making waste disposal costly **(Srivastava et al., 1985).**

Buttermilk and separated milk are obtained during the manufacture of butter and cream respectively. Buttermilk contains almost all other milk solids except fat and is produced while butter and ghee are manufactured. Buttermilk is a good source of proteins, carbohydrates, minerals, vitamin-A tocopherol and cholesterol.

Due to urbanization and advancement in technology, the food habits of people are changing. The trend is more towards the convenience foods. Therefore, the utilization of milk byproducts in bakery products will improve the nutritive value of these products at a cheaper rate

Keeping in view the above fact the present work dealt with the following objectives:-

- To study the chemical composition of raw material such as wheat flour, buttermilk etc.
- To study the effect of Incorporation levels of buttermilk on the physical characteristics such as mixing, dough handling etc.
- To study the effect of incorporation of buttermilk on sensory characteristics of cookies

Material and Method

Atta (whole wheat flour) was procured from the local market of Jhansi(U.P.) and were analyzed for moisture, ash, crude protein, crude fat, crude fiber and nitrogen free extract **(AOAC, 2000).**

Preparation of Buttermilk

Buttermilk was prepared in the laboratory as a by-product of butter preparation. Cream procured from local market was pasteurized at 71°C for 20 minutes, cooled immediately and stored overnight in refrigerator. Cream was then churned in a butter churn to obtain buttermilk. This buttermilk was stored in a refrigerator for further use.

Prepare the cream
↓
Churning of the cream
↓
Add small quantity of water
↓
Remove the butter or cream from the above
↓
Remaining liquid left is buttermilk

Figure 27.1: *Preparation of Buttermilk*

Chemical analysis: Buttermilk and Wheat flour were analyzed for fat, moisture, ash, total soluble solids, pH and acidity **(AOAC, 2000).**

TABLE 27.1

Chemical Composition of Buttermilk and Wheat Flour

Chemical Constituents (%)	*Buttermilk*	*Wheat Flour*
Moisture	90.00	12.09
Protein	3.98	10.42
Fat	0.67	0.30
Ash	0.32	0.27
Carbohydrate (by difference)	5.03	75.33
Total solids	10.0	86.32
Minerals	------	0.67

TABLE 27.2

Mineral Content of Buttermilk

Minerals (mg/100 g)	*Amount in Buttermilk*
Calcium	406.81
Iron	3.42
Potassium	265.82
Magnesium	67.31
Phosphorus	382.11
Sodium	0.12

Cookies: The control dough was mixed using optimum level of water. Water was replaced at 20 to 100 levels with buttermilk in test sample. The dough was mixed, sheeted and cut into circular cookies of diameter 4.5 cm and baked for 12 minute at 380°F. For preparation of cookies following method had been used.

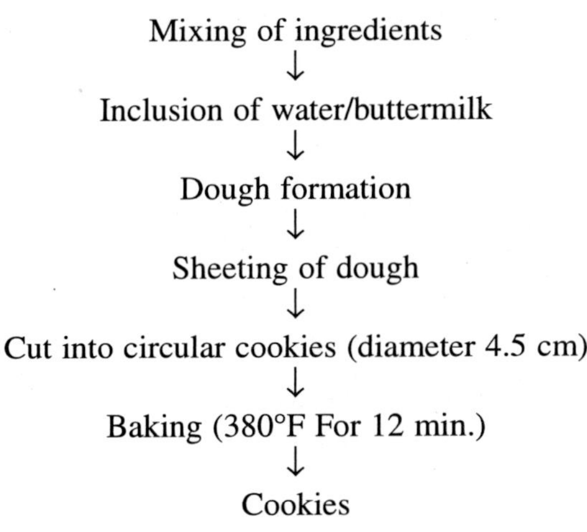

Figure 27.2: *Preparation of Cookies*

For baking cookies AACC (1990) method was followed using following ingredients:

TABLE 27.3

Ingredients used in Cookies Making

Ingredients	*Quality(g)*
Flour	100.0
Sugar	57.8
Bakery shortening	28.44
Salt	0.993
Dextrose	0.878
Sodium bicarbonate	1.1
Water	Optimum (ml)

Sensory Analysis

Sensory evaluation of cookies was carried out by a panel of semi-trained judges on a point hedonic scale. Cookies were scored for texture, flavour, mouthfeel and overall acceptability. The data obtained were analysed statistically on a completely randomized design using analysis of variance technique **(Snedecor and Cochran,1968)** to find if the differences were significant or not.

Results and Discussion

The chemical composition of buttermilk showed that it contained 10 percent total solids. The protein, fat and ash content of buttermilk were 3.98, 0.67 and 0.32 percent, respectively (Table 27.3). The values are similar to those as reported by **Nutting et al.,(1970)**.The wheat flour contained 86.32 percent total solids. The protein, fat, ash and carbohydrate content were 10.42, 0.30, 0.27and 75.33 percent, respectively (Table 27.3). Values of these constituents of wheat flour was in the same range as reported earlier for Indian wheat's by **Hira et al.,(1991), C. Gopalan et al.,(1998)**.

The mineral content varied appreciably in buttermilk. Calcium content of buttermilk was 406.81mg/100g. Iron content of buttermilk was quite low being 3.42.Potassium, magnesium, phosphorus and sodium was also found to be low in buttermilk i.e. 265.82, 67.31, 0.12 and 382.11 mg/100g respectively (Table 27.2)

The average estimates of water replacement with buttermilk for fat, protein, ash, and free fatty acids (FFA) are given in table 27.4)

TABLE 27.4

Effect of Water Replacement with Buttermilk on Proximate Components of Cookies

Replacement %	*Fat%*	*Protein%*	*Ash%*	*FFA%*
0	29.00	9.79	0.705	0.011
20	29.50	9.88	0.708	0.034
40	29.56	9.91	0.743	0.031
60	29.66	9.94	0.856	0.027
80	29.71	10.00	0.916	0.025
100	29.81	10.05	0.918	0.021
Mean	**29.54**	**9.93**	**0.81**	**0.06**
CV	**0.97**	**0.92**	**12.5**	**32.6**

As inferred from Table 27.5, width of cookies increased significantly ($p < 0.05$) with enhanced level of buttermilk, thereby, significantly improving the spread ratio of cookies. Spread ratio was found to be 5.66 in control and increased to 6.10 in 100 percent buttermilk incorporated cookies. The increased in width of cookies with incorporation of buttermilk might be due to the additional phospholipids content acting as emulsifier.

TABLE 27.5

Effect of Water Replacement with Buttermilk on Spread Ratio of Cookies

Replacement %	*Width (cm)*	*Thickness (cm)*	*Spread Ratio*
0	6.01	1.06	5.66
20	6.06	1.06	5.70
40	6.18	1.06	5.84
60	6.28	1.05	5.98
80	6.23	1.03	6.06
100	6.25	1.02	6.10
Mean	**6.17**	**1.04**	**5.89**
CV	**5.61**	**1.66**	**3.13**

Sensory Characteristics

TABLE 27.6

Effect of Water Replacement with Buttermilk on Organoleptic Scores of Cookies

Replacement %	*Appearance*	*Colour*	*Texture*	*Flavour*	*Mouthfeel*	*Overall Acceptability*
0	6.83	7.16	6.83	7.17	6.83	7.11
20	6.95	7.45	7.60	7.45	7.45	7.32
40	7.20	7.55	7.82	7.70	7.60	7.68
60	7.35	7.85	7.95	7.88	7.88	7.76
80	7.55	8.20	8.20	7.55	7.88	7.90
100	7.30	7.66	8.00	7.66	8.00	7.71
Mean	**7.19**	**7.64**	**7.73**	**7.56**	**7.60**	**7.58**
CV	**3.67**	**4.64**	**6.27**	**3.21**	**5.67**	**3.95**

Effect of various levels of buttermilk on sensory attributes of appearance, color, texture, flavor, mouthfeel and overall acceptability of cookies was found to be non significant except for texture and mouth feel, which were affected significantly (Table 27.6). Crispiness of cookies improved with increase in levels of buttermilk, thereby, resulting in improvement of sensory scores for the attributes of texture. This was in accordance with the study of **Cooper et al., (1984)** who reported that as the level of the milk proteins increased, the mouth feel tended to become finer in character and the color become finer in character and the color become brown due to the millard browning reactions between the protein and lactose present.

References

Srivastava M.K., Trimurthulu N. and Lohani P.P., (1985). Utilization of byproducts of dairy industries—whey and buttermilk. Indian dairyman 37(11) 507.

AOAC, (2000). Officials Methods of Analysis. Association of Official Analytical Chemists, Arlington, D.C., U.S.A.

Cooper H.R., Pattern J.C. and Fletcher R.H., (1984). Effect of some insoluble milk proteins on sensory characteristics of appearance and texture in cookies. J Fd Sci 49(2): L 376-79.

Snedecor, G.W. and Cochran, *Statistical Methods*, 6th ed. Ames Iowa State Press, Iowa, U.S.A.

Hira, C.K., Sadana, B.K. and Grover, K. 1991. Food consumption pattern of farm families in rice wheat cropping pattern. All India Co-ordinate research project in M.Sc. Foods and Nutrition unit, Punjab Agricultural University, Ludhiana, India.

Gopalan, C., Rama Sastari, B.V. and Balasubramanium, S.E. 1985. *Nutritive Value of Indian Foods.* National Institute of Nutrition, ICMR, Hyderabad, India.

28

Characterization of Steamed Idli on the Basis of Physical Properties

K. Prasad* and T. Bhatia

Abstract

Idli is a fermented cereal based snack food that is widely consumed in southern parts of Indian subcontinent. The different physical properties that included dimensional, gravimetric, frictional and optical properties were evaluated. The unit mass of lentils based functional idli and chickpea based functional idli were in the range of 21.84 ± 0.54 to 32.31 ± 2.88 g and 26.82 ± 2.72 to 34.27 ± 3.12 g, respectively. Longitudinal dimension (L) of the idli made from the ratio of lentils/black gram ranged from 5.79 ± 0.46 to 6.41 ± 0.10 cm whereas for the idli made from chickpea splits/ black gram ranged from 6.12 ± 0.03 to 6.59 ± 0.10 cm. For the transverse dimensions (W) the idli made from lentils/black gram ratio was in the range of 5.48 ± 0.37 to 6.09 ± 0.16 cm, whereas, in the other case, it ranged from 5.74 ± 0.25 to 6.28 ± 0.14 cm. The sphericity of the lentils based idli is in the range of 61.61 ± 0.79 – 69.95 ± 0.38% whereas, for the other case it lies in the range of 65.45 ± 0.34 – 70.24 ± 1.72%. The aspect ratio of lentils based idli and chickpea based idli were in the range of 89.11 ± 3.22 to 97.39 ± 1.77% and 87.05 ± 3.66 to 97.71 ± 1.56%, respectively. The true density and bulk density of the lentils based idli ranged from 0.62 ± 0.02 to 0.77 ± 0.01 g/ml and 0.33 ± 0.01 to 0.43 ± 0.04 g/ml, respectively. On the other hand, the idli made from chickpea/black gram ranged from 0.63 ± 0.01 to 0.78 ± 0.01 g/ml and 0.36 ± 0.01 to 0.46 ± 0.01 g/ml, respectively. The ΔE of lentils made idli ranged from 3.75 to 7.82 and those of chickpea made idli ranged from 3.34 to 7.87.

Keywords: *Idly, fermented product, physical property, colour, RSM, optimization*

Introduction

The idli also romanized “idly” or “iddly”, is a savory cake (Wikipedia, the encyclopedia). Idli is a very popular fermented breakfast food staple consumed in the Indian subcontinent, especially in southern parts. The major ingredients are rice (*Oryza sativum*) and black gram (*Phaseolus mungo*) in the ratio of 2:1

to 3:1 (Nagaraju and Manohar, 2000). It is famous for its soft, spongy texture, desirable sour taste and characteristic aroma. It resembles steamed, sour dough bread (Steinkraus *et al.*, 1967). Idli has a circular shape of approximately 7 – 10 cm diameter (depending on the mold size), flat with convex lower and upper surface, so that the product is thick at the center (2 – 3 cm) and tapering towards periphery (Desikacharr *et al.*, 1960). Idli is considered as a healthy food because fermentation process breaks down the starches present so that it is more easily metabolized by the body.

Guar gum is a neutral, water soluble, galactomannan, extracted from the seeds of *Cyamopsis tetragonoloba* (Maier *et al.,* 1993). The guar seeds are dehusked, milled and screened to obtain the guar gum. It consists of the linear backbone of β- (1$\rightarrow$ 4) - linked D-mannose units (M) and is solubilized by the presence of randomly attached α-(1 $\rightarrow$ 6) – linked galactose units as side chains (Mc Cleary *et al.*, 1985).

Guar gum has a range of nutritional and medicinal effects which make it a very popular product to be used in our daily life. Guar gum is water soluble fiber that acts as a bulk forming laxative and it is claimed to be effective in promoting regular bowel movements and relieve constipation and chronic related functional bowel ailments such as diverticulosis, Crohn's disease, colitis and irritable bowel syndrome, among others. The increased mass in the intestines stimulates the movement of waste and toxins from the system, which is particularly helpful for good colon health, because it speeds the removal of waste and bacteria from the bowel and colon. In addition, because it is soluble, it is also able to absorb toxic substances (bacteria) that cause infective diarrhea (Wikipedia, the encyclopedia). Guar gum has been considered of interest with regards to both weight loss and diabetic diets. It is a thermogenic substance. Moreover, its low digestibility lends its use in recipes as filler, which can help to provide satiety, or slow the digestion of a meal, thus lowering the glycemic index of that meal (Brown and Livesey, 1994).

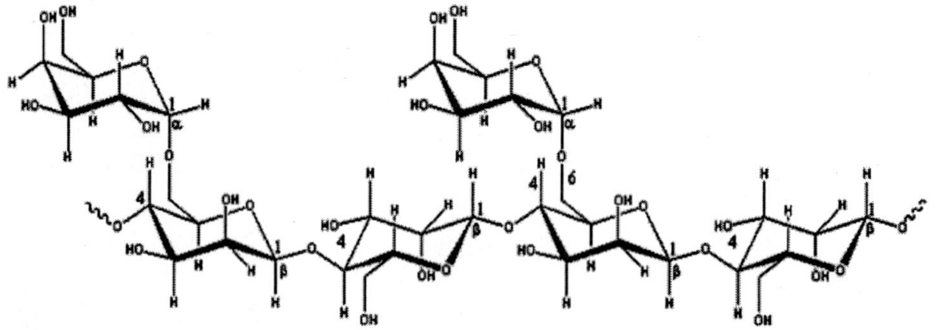

Figure 28.1: *Structure of Guar Gum*

Guar gum has a range of nutritional and medicinal effects which make it a very popular product to be used in our daily life. Guar gum is water soluble fiber that acts as a bulk forming laxative and it is claimed to be effective in promoting regular bowel movements and relieve constipation and chronic related functional bowel ailments such as diverticulosis, Crohn's disease, colitis and irritable bowel syndrome, among others. The increased mass in the intestines stimulates the movement of waste and toxins from the system, which is particularly helpful for good colon health, because it speeds the removal of waste and bacteria from the bowel and colon. In addition, because it is soluble, it is also able to absorb toxic substances (bacteria) that cause infective diarrhea (Wikipedia, the encyclopedia). Guar gum has been considered of interest with regards to both weight loss and diabetic diets. It is a thermogenic substance. Moreover, its low digestibility lends its use in recipes as filler, which can help to provide satiety, or slow the digestion of a meal, thus lowering the glycemic index of that meal (Brown and Livesey, 1994).

Studies have been done on Rheology and particle size changes during idli fermentation by Nagaraju and Manohar, 2000. Effect of α-amylase addition on fermentation of idli- a popular south Indian cereal-legume based snack food has been investigated by Iyer and Ananthanarayan, 2008. Investigations have been done on thermal diffusivity of idli batter (Murthy and Rao, 1997). Studies have been done on the effect of stabilizers on stabilization of idli (traditional south Indian food) batter during storage by Nisha *et al.,* 2005. However, no results for the physical properties of idli yet appear to be available.

The objective of the present study was to portray the idli made from lentils and chickpea splits on the basis of physical properties mainly the gravimetric, frictional, dimensional and optical properties.

Materials and Methods

Raw materials: Rice (Permal variety) and decuticled black gram were procured from the local market of Sangrur, Punjab. They were cleaned in an air classifier to remove lighter foreign matter such as dust, dirt, chaff, immature and broken splits. Edible grade salt purchased from the local market was used. Guar gum was arranged from one of the laboratories of Sant Longowal Institute of Engineering and Technology, Sangrur.

Design of Experiment: Response surface methodology (RSM) is a useful technique for studying the effect of several factors influencing the response by varying them simultaneously. For the optimization of process conditions, the experiments were conducted according to second order Central Composite Rotatable Design (CCRD).

The ratio of chickpea or lentils/black gram splits and hydrocolloids concentration was the process variables used for the making of the idli (Table 25.1). For the two factors, this design was made up of a full 2^2 factorial design with five replications of the centre points (all factors at level 0). A set of 13 experiments were carried out as per explained below:

Total No. of experiments = $2^{\text{No. of variables}}$ + 2 × No. of variables + Central point experiments.

Hence, for two variables,

Total number of experiments = $2^2 + 2 \times 2 + 5 = 13$

Five different levels for each experiment in coded form are -1.5, – 0.75, 0, 0.75, and 1.5.

The relationship between the coded and un-coded form of the variables is

$$\text{Coded Value} = \frac{(\text{Un Coded Value} - \text{Central Value})}{(\text{Interval between Two Successive Intervals})}$$

The range and levels of experimental variables investigated are presented in the table 28.1. The central value (zero level) chosen for experimental design were: ratio of chickpea or lentils/black gram 0.4 and hydrocolloids 1 g (Table 28.1).

Physical Properties: The shape of the idli was found to be hemisphere with three major perpendicular dimensions, length (L), width (W) and thickness (T). The physical dimensions were determined randomly measuring the length, width and thickness of idli using dial type vernier caliper (Mitutoyo Corporation, Japan) having least count 0.02mm.

The geometric mean dimension (D_e) of idli was found using the relationship given by Mohsenin (1970) as in Eq. (1).

$$D_e = (LWT)^{1/3} \tag{1}$$

The criteria used to describe the shape of the idli are the sphericity and aspect ratio. Thus, the sphericity (S_p) was accordingly computed (Mohsenin, 1970) as in Eq. (2).

$$S_P = \frac{(LWT)^{1/3}}{L} \times 100 \tag{2}$$

The aspect ratio (R_a) was calculated (Maduako & Faborode, 1990) as

$$R_a = \frac{W}{L} \times 100 \tag{3}$$

The surface area (S_a) of idli as semi sphere was calculated using the relationship given by McCabe *et al.* (1986).

$$S_a = \pi D_e^2 \tag{4}$$

The weights of the idli were recorded using electronic balance (Ishida Co. Ltd., Japan) to an accuracy of 0.001 g. The true density of an idli is defined as the ratio of mass of idli to the solid volume occupied (Deshpande *et al.,* 1993). The true density was determined by crushing the idli properly and taking its volume in a measuring cylinder and after that taking the weight of the crushed idli. The bulk density (g/cm3) of the samples was measured by the seed displacement method using mustard seed as in case of bread (Nisha *et al.*, 2005). The bulk density was calculated using the following formula:

Specific volume = [(Initial Vol.) – (Displacement Vol.)]/ Wt of the products (5)

Bulk density = 1/ Specific volume (6)

The porosity (ε) of idli was computed from the values of true density (ρ_t) and bulk density (ρ_b) using the relationship given by Mohsenin (1970)

$$\varepsilon = \frac{\rho_t - \rho_b}{\rho_t} \times 100 \tag{7}$$

Results and Discussions

The summary of the results for physical parameters are shown in Table 28.2 and Table 28.3. The graphical representations of some of the physical properties are shown in Fig. 28.2 and Fig. 28.3. Longitudinal dimension (L) of the idli made from the ratio of lentils/black gram ranged from 5.79 ± 0.46 to 6.41 ± 0.10 cm (Table 2) whereas, the length (L) of the idli made from chickpea splits/ black gram ranged from 6.12 ± 0.03 to 6.59 ± 0.10 cm (Table 28.3). For the transverse dimensions (W) the idli made from lentils/black gram ratio was in the range of 5.48 ± 0.37 to 6.09 ± 0.16 cm, whereas, in the other case, it ranged from 5.74 ± 0.25 to 6.28 ± 0.14 cm. The range of thickness (T) of the idli made from the lentils/ black gram or chickpea/ black gram is illustrated in Table 28.2 and Table 28.3, respectively. It has been tried to make a clear view of the range for the sphericity and the aspect ratio in the Fig. 28.2. It is seen that sphericity of the idli made from lentils/black gram ratio lies in the range of 61.61 ± 0.79 – 69.95 ± 0.38% whereas, for the other case it lies in the range of 65.45 ± 0.34 – 70.24 ± 1.72%. The aspect ratio for the former case is in the range of 89.11 ± 3.22 to 97.39 ± 1.77%, whereas, for the latter case, it lies in the range of 87.05 ± 3.66 to 97.71 ± 1.56%. The range of the surface area has been cited in the graphical diagrams of Fig. 28.2 and 28.3. The true density and the bulk density of the idli made from lentils/black gram ratio lies in the range of 0.62 ± 0.02 to 0.77 ± 0.01 g/ml and 0.33 ± 0.01 to 0.43 ± 0.04 g/ml, respectively. On the other hand, the idli made from chickpea/black gram ranged from 0.63 ± 0.01 to 0.78 ± 0.01 g/ml and 0.36 ± 0.01 to 0.46 ± 0.01 g/ml, respectively. The porosity for the former case is in the range of 36.48 ± 1.83 to 50.83 ± 0.86% and in the latter case it is in the range of 32.65 ± 1.65 to 50.33 ± 1.19%.

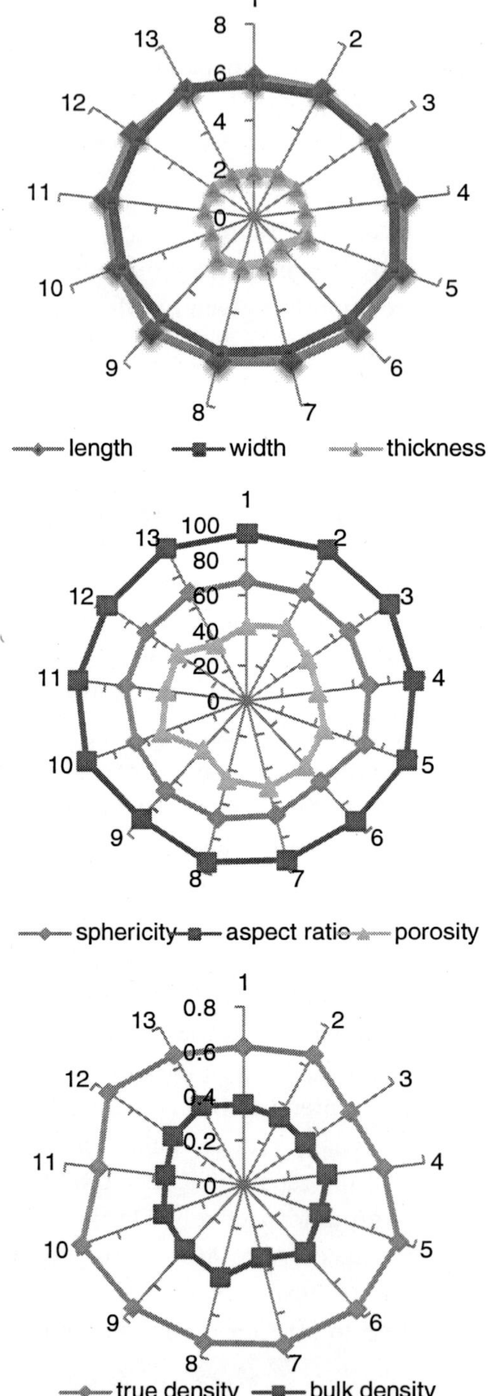

Figure. 28.2: *Radar Representation of Lentils made Functional Idli*

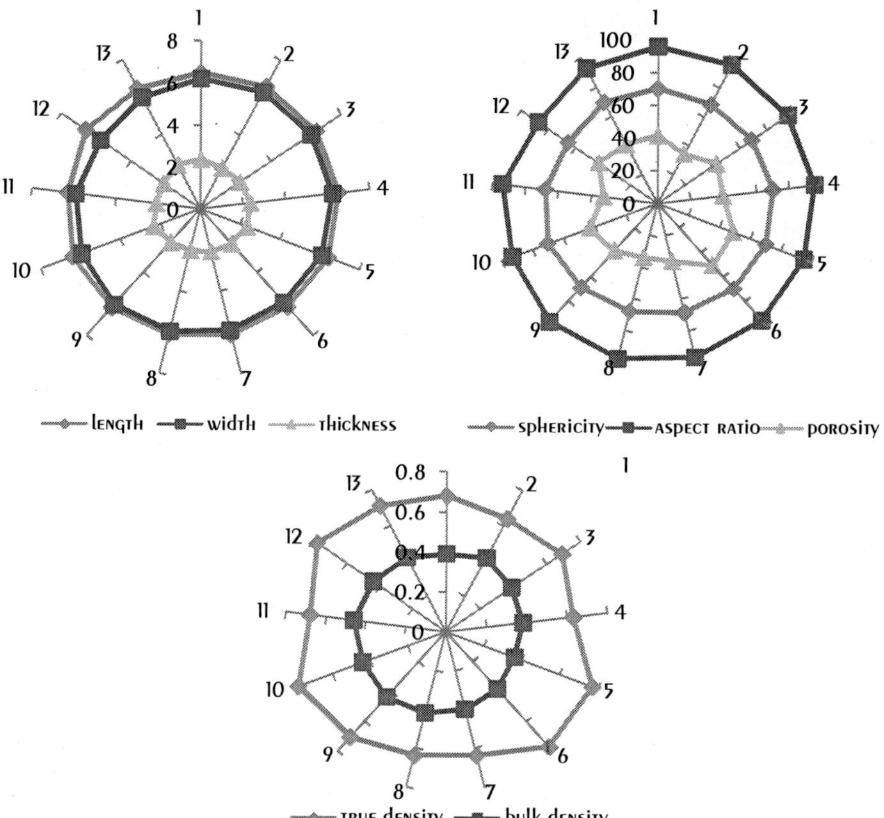

Figure. 28.3: *Radar Representation of Lentils made Functional Idli*

TABLE 28.1

Actual Values of Independent Variables at Five Levels of the CCRD Design

Independent Variables	*Unit*	*Symbol*	*Levels in coded form*				
			– 1.5	**– 0.75**	**0**	**+ 0.75**	**+ 1.5**
CP or lentils/BG		X_1	0	0.12	0.4	0.68	0.8
Hydrocolloids	g	X_2	0	0.29	1	1.71	2

The L, a & b values of idli made from lentils ranged from 76.839 to 79.788, –0.865 to 0.166 and 14.573 to 21.661, respectively (Table 28.4), while the L, a & b values for idli made from chickpea ranged from 74.382 to 79.676, –0.175 to 1.601 and 13.522 to 18.26, respectively (Table 5). The "E of lentils made idli ranged from 3.75 to 7.82 (Table 28.4) and those of chickpea made idli ranged from 3.34 to 7.87 (Table 28.5).

TABLE 28.2

Physical Properties of Lentils Made Functional Idli

S. No✱	*Length (cm)*	*Width (cm)*	*Thickness (cm)*	*Surface area (cm^2)*	*GM Dia** *(cm)*	*Weight (g)*	*Sphericity (%)*	*Aspect ratio (%)*	*True density (g/ml)*	*Bulk density (g/ml)*	*Porosity (%)*
1	5.79±0.46	5.48±0.37	1.87±0.11	48.21±5.41	3.91±0.22	21.84±0.54	67.7±2.73	94.76±4.04	0.62±0.02	0.36±0.02	42.77±2.45
2	5.91±0.07	5.72±0.03	2.02±0.07	52.31±2.01	4.08±0.08	24.67±2.67	68.95±0.48	96.62±0.73	0.66±0.06	0.34±0.02	47.16±0.45
3	5.97±0.16	5.75±0.10	2.08±0.05	53.84±1.82	3.85±0.50	25.52±2.66	69.14±0.98	96.17±1.82	0.57±0.002	0.33±0.01	41.22±1.84
4	6.14±0.14	5.73±0.04	2.09±0.13	55.08±2.41	4.18±0.09	25.01±2.67	68.39±1.97	93.41±2.57	0.62±0.02	0.37±0.01	39.89±0.21
5	6.41±0.10	6.09±0.16	2.32±0.016	63.26±2.05	4.48±0.07	29.96±1.96	69.95±0.38	94.95±1.32	0.73±0.05	0.36±0.02	45.94±4.97
6	6.3±0.36	5.73±0.1	1.62±0.12	47.38±4.75	3.88±0.19	29.60±3.32	61.61±0.79	91.15±3.99	0.75±0.02	0.41±0.02	49.07±4.26
7	6.13±0.11	5.68±0.12	1.97±0.13	52.58±2.57	4.09±0.1	28.75±1.14	66.77±1.87	92.81±0.92	0.74±0.007	0.34±0.01	50.41±7.10
8	6.13±0.03	5.75±0.11	2.11±0.19	55.53±2.58	4.20±0.09	30.86±5.85	68.63±1.71	93.91±1.83	0.73±0.04	0.43±0.04	46.01±4.67
9	6.34±0.25	5.65±0.05	2.28±0.07	59.19±2.39	4.34±0.08	32.31±2.88	68.42±1.62	89.11±3.22	0.74±0.02	0.39±0.05	36.84±2.72
10	6.00±0.19	5.77±0.15	1.85±0.12	64.01±1.46	3.99±0.14	31.69±0.61	66.53±1.98	96.09±2.64	0.77±0.01	0.38±0.01	50.83±0.86
11	6.11±0.1	5.84±0.05	2.04±0.04	57.56±0.97	4.18±0.05	29.97±1.94	68.34±0.88	95.54±0.87	0.65±0.01	0.35±0.01	45.83±2.39
12	6.03±0.07	5.74±0.11	2.03±0.07	58.46±1.23	4.12±0.04	31.51±2.75	68.29±1.21	95.21±2.70	0.73±0.02		46.87±0.44
13	5.91±0.17	6.02±0.02	1.93±0.07	63.44±1.73	4.09±0.05	30.14±0.39	69.29±1.96	97.39±1.77	0.66±0.02	0.40±0.02	36.48±1.83

« Geometric mean diameter

- All readings are mean ± Standard deviation

TABLE 28.3

Physical Properties of Chickpea Made Functional Idli

S. No+	*Length (cm)*	*Width (cm)*	*Thickness (cm)*	*Surface area (cm^2)*	*G.M. Dia* (cm)*	*Weight (g)*	*Sphericity (%)*	*Aspect ratio (%)*	*True Density (g/ml)*	*Bulk density (g/ml)*	*Porosity (%)*
1	6.48±0.07	6.2±0.1	2.35±0.14	65.15±2.01	4.55±0.07	27.02±2.01	70.24±1.72	95.64±2.17	0.68±0.01	0.39±0.01	41.62±0.87
2	6.56±0.29	6.27±0.21	2.2±0.22	63.11±4.38	4.48±0.16	29.28±3.24	68.32±3.11	95.56±1.36	0.64±0.01	0.42±0.01	34.05±1.31
3	6.56±0.05	6.28±0.14	2.24±0.03	63.91±1.50	4.51±0.05	31.49±3.26	68.78±0.65	95.40±0.11	0.69±0.01	0.39±0.01	43.09±1.29
4	6.51±0.17	6.23±0.15	2.35±0.08	65.46±3.44	4.56±0.12	26.82±2.72	70.13±0.94	95.80±0.67	0.63±0.01	0.38±0.01	39.81±0.81
5	6.41±0.10	6.09±0.16	2.32±0.06	63.26±2.05	4.48±0.07	29.96±1.96	69.96±0.38	94.95±1.32	0.77±0.01	0.36±0.01	48.98±7.64
6	6.20±0.26	5.91±0.17	2.2±0.1	58.59±4.11	4.32±0.15	28.69±3.80	69.63±0.65	95.42±1.98	0.77±0.02	0.38±0.02	50.33±1.19
7	6.13±0.02	5.92±0.07	2.08±0.12	56.06±2.60	4.22±0.09	34.27±3.12	68.93±1.64	96.68±1.11	0.64±0.01	0.41±0.01	36.55±0.35
8	6.12±0.03	5.97±0.07	2.01±0.05	54.90±1.38	4.18±0.05	33.66±1.82	68.34±1.18	97.71±1.56	0.64±0.01	0.42±0.01	34.01±1.71
9	6.27±0.09	6.08±0.16	2.13±0.26	58.92±6.08	4.33±0.22	32.73±1.37	68.94±2.46	96.91±1.39	0.71±0.01	0.44±0.01	38.65±0.60
10	6.43±0.15	5.98±0.1	2.36±0.06	50.14±3.46	4.51±0.51	30.59±1.49	70.21±1.25	93.07±1.70	0.78±0.01	0.44±0.02	43.65±2.19
11	6.32±0.05	5.93±0.06	2.11±0.04	54.90±1.20	4.28±0.04	31.91±0.66	67.72±0.26	93.83±0.58	0.68±0.02	0.46±0.01	32.65±1.65
12	6.59±0.10	5.74±0.25	2.13±0.09	53.33±0.93	4.31±0.04	28.64±1.35	65.45±0.34	87.05±3.66	0.78±0.01	0.44±0.01	43.29±1.15
13	6.45±0.08	5.99±0.12	2.35±0.11	52.65±1.21	4.49±0.06	32.55±1.59	69.67±1.27	92.86±0.87	0.71±0.01	0.42±0.01	40.62±2.05

* Geometric mean diameter

+ All readings are mean ± standard deviation

TABLE 28.4

Optical Characteristics* of Lentils Made Functional Idli

Exp. No.	*Optical parameters*			
	L	*a*	*b*	Δ*E*
1	79.676	–0.715	14.673	3.75
2	78.472	–0.356	17	4.49
3	77.944	0.166	14.573	3.81
4	79.167	–0.864	16.726	4.50
5	79.676	–0.715	14.673	3.75
6	76.839	–0.266	17.007	5.39
7	78.571	–0.203	19.572	6.08
8	78.524	–0.527	15.639	4.11
9	78.066	–0.28	20.748	7.25
10	78.876	–0.274	21.661	7.80
11	79.362	–0.287	20.471	6.67
12	78.256	–0.321	21.481	7.82
13	79.788	–0.276	21.222	7.26

*Obtained from hunter color technique.

TABLE 28.5

Optical Characteristics* of Chickpea Made Idli

Exp. No.	*Optical parameters*			
	L	*a*	*b*	Δ*E*
1	76.423	0.335	18.26	6.01
2	76.806	1.248	15.665	4.21
3	77.869	0.956	15.036	3.34
4	76.115	1.133	15.787	4.89
5	79.676	–0.715	14.673	3.76
6	76.087	0.92	15.889	5.03
7	75.533	1.601	14.119	5.23
8	72.963	1.017	15.626	7.87
9	74.638	0.371	13.522	6.56
10	75.142	0.412	13.867	6.03
11	75.112	0.354	13.993	6.04
12	75.729	0.422	13.988	5.49
13	74.382	0.456	14.181	6.67

*Obtained from hunter color technique.

Conclusion

The average values of physical properties of idli made from lentils/black gram ratio and chickpea/black gram ratio length (L), width (W), thickness (T), unit mass were in the range of 5.79 to 6.41 cm and 6.12 to 6.59 cm, 5.48 to 6.09 cm and 5.74 to 6.28 cm, 1.62 to 2.32 cm and 2.01 to 2.36 cm, 21.84 to 32.31 g and 26.82 to 34.27 g, respectively. The calculated physical properties like geometric mean diameter, surface area, porosity, true density, and aspect ratio for the idli made from lentils/ black gram and chickpea/ black gram lied in the range of 3.85 to 4.48 cm and 4.18 to 4.56 cm, 47.38 to 64.01 cm^2 and 50.14 to 65.46 cm^2, 36.48 to 50.83% and 32.65 to 50.33%, 0.62 to 0.77 g/ml and 0.63 – 0.78 g/ml, 89.11 to 97.39% and 87.05 to 97.71%, respectively. The L, a & b values of idli made from lentils ranged from 76.839 to 79.788, –0.865 to 0.166 and 14.573 to 21.661, respectively, while the L, a & b values for idli made from chickpea ranged from 74.382 to 79.676, –0.175 to 1.601 and 13.522 to 18.26, respectively All standard deviation of all the measured parameters ranged between 0.004 and 7.10 showing near uniform dispersion about their respective mean values.

References

Brown, J., and Livesey, G. (1994). Energy balance and expenditure while consuming guar gum at various fat intakes and ambient temperatures. American Journal of Clinical Nutrition, 60(6), 956-964.

Deshpande, S.D., Bal, S. and Ojha, T.P. (1993). Physical properties of soybean. Journal of Agricultural Engineering Research, 39, 259-268.

Desikacharr, H.S.R., Radhakrishnamurthy, R., Rao, G.R., Kadol, S.B., Srinivasan, M. and Subramanyan, V. (1960). Studies on Idli fermentation. Part 1. Some accompanying changes in the batter. Journal of Scientific Industrial Research (India), 19C, 168-172.

Iyer Bharti, K. and Ananthanarayan, A. (2008). Effect of α-amylase addition on the fermentation of idli-a popular south Indian cereal- legume based snack food. Lebensmittel-Wissenschaft Technology, 41, 1053-1059.

Maduako, J.N. and Faborode, M.O. (1990). Some physical properties of cocoa podsin relation to primary processing. Ife Journal of Technology, 2, 1-7.

Maier, H., Anderson, M., Karl, C., Magnuson, K. and Whistler, R.L. (1993). Guar, locust bean gum, tara, and fenugreek gums. In R.L. Whistler, & J.N. BeMiller (Eds.), Industrial gums (3rd ed.). New York: Academic Press, 181–226.

McCabe, W.L., Smith, J.C. and Harriorth, P. (1986). Unit operations of chemical engineering. New York: McGraw-Hill Book Company, p-1152 .

McCleary, B.V., Clark, A.H., Dea, I.C.M. and Rees, D.A. (1985). The fine structures of carob and guar galactomannans. Carbohydrate Research, 139, 237–260.

Mohsenin, N. N. (1970). Physical properties of plant and animal materials. Vol.1. Physical characteristics and mechanical properties, Gordon and Breach Science Publishers, New York, U.S.A, p-408.

Murthy, C.T. and Rao, P.N.S. (1997). Thermal diffusivity of idli batter. Journal of Food Engineering, 33, 299-304.

Nagaraju, V.D. and Manohar, B. (2000). Rheology and particle size changes during Idli fermentation, 43, 167-171.

Nisha, P., Ananthanarayan, L. and Singhal, R.S. (2005). Effect of stabilizers on stabilization of idli (traditional south Indian food) batter during storage. Food Hydrocolloids, 19(2), 179–186.

Steinkraus, K.H., Van Veen, A.G. and Thiebeau, D.B. (1967). Studies on Idli - an Indian fermented black gram - rice food. Food Technology, 21, 91-919.

Wikipedia, the encyclopedia (2009). http://www.wikipedia.com.